Structural and Stress Ana

Also available from Taylor & Francis

Design of Structural Elements 2nd edition
C. Arya

Hb: ISBN 9780415268448
Pb: ISBN 9780415268455

Examples in Structural Analysis
W.M.C. McKenzie

Hb: ISBN 9780415370530
Pb: ISBN 9780415370547

Structural Analysis 5th edition
A. Ghali, A. Neville and T.G. Brown

Hb: ISBN 9780415280914
Pb: ISBN 9780415280921

Structures: From Theory to Practice
A. Jennings

Hb: ISBN 9780415268424
Pb: ISBN 9780415268431

Information and ordering details

For price availability and ordering visit our website **www.tandfbuiltenvironment.com/**

Alternatively our books are available from all good bookshops.

Structural and Stress Analysis

Theories, tutorials and examples

Jianqiao Ye

Taylor & Francis
Taylor & Francis Group

LONDON AND NEW YORK

First published 2008
by Taylor & Francis
2 Park Square, Milton Park, Abingdon, Oxon OX14 4RN

Simultaneously published in the USA and Canada
by Taylor & Francis
270 Madison Ave, New York, NY 10016, USA

Taylor & Francis is an imprint of the
Taylor & Francis Group, an informa business

© 2008 Jianqiao Ye

Typeset in Frutiger Light by
Integra Software Services Pvt. Ltd., Pondicherry, India
Printed and bound in Great Britain by
TJ International, Padstow, Cornwall

British Library Cataloguing in Publication Data
A catalogue record for this book is available from the British Library

Library of Congress Cataloging in Publication Data
Ye, Jianqiao, 1957-
Structural and stress analysis : theories, tutorials and examples / Jianqiao Ye.
p. cm.
Includes bibliographical references and index.
ISBN 978-0-415-36865-0 (hardback : alk. paper) --
ISBN 978-0-415-36879-7 (paperback : alk.
paper) -- ISBN 978-0-203-02900-8 (ebook) 1. Structural
analysis (Engineering) I. Title.
TA645.Y43 2008
624.1'71--dc22
2007030738

ISBN10: 0–415–36865–0 (hbk)
ISBN10: 0–415–36879–0 (pbk)
ISBN10: 0–203–02900–3 (ebk)

ISBN13: 978–0–415–36865–0 (hbk)
ISBN13: 978–0–415–36879–7 (pbk)
ISBN13: 978–0–203–02900–8 (ebk)

To my wife Qin and my daughter Helen
with love and gratitude

Contents

Preface

This book is not intended to be an additional textbook of structural and stress analysis for students who have already been offered many excellent textbooks which are available on the market. Instead of going through rigorous coverage of the mathematics and theories, this book summarizes major concepts and important points that should be fully understood before students claim that they have successfully completed the subject. One of the main features of this book is that it aims at helping students to understand the subject through asking and answering conceptual questions, in addition to solving problems based on applying the derived formulas.

It has been found that by the end of a Structural and Stress Analysis course, most of our students can follow the instructions given by their lecturers and can solve problems if they can identify suitable formulas. However, they may not necessarily fully understand what they are trying to solve and what is really meant by the solution they have obtained. For example, they may have found the correct value of a stress, but may not understand what is meant by "stress". They may be able to find the direction of a principal stress if they know the formula, but may not be able to give a rough prediction of the direction without carrying out a calculation. To address these issues, understanding all the important concepts of structures and stresses is essential. Unfortunately, this has not been appropriately highlighted in the mainstream textbooks since the ultimate task of these textbooks is to establish the fundamental theories of the subject and to show the students how to derive and use the formulas.

Leaving out all the detailed mathematics and theories found in textbooks, each chapter of this book begins with a summary of key issues and relevant formulas. This is followed by a key points review to identify important concepts that are essential for students' understanding of the chapter. Next, numerical examples are used to illustrate these concepts and the application of the formulas. A short discussion of the problem is always provided before following the solution procedure to make sure that students know not only how but also why a formula should be used in such a way. Unlike most of the textbooks available on the market, this book asks students to answer only questions that require minimum or no numerical calculations. Questions requiring extensive numerical calculations are not duplicated here since they can be easily found from other textbooks. The conceptual questions ask students to review important concepts and test their understanding of the concepts. These questions can also be used by lecturers to organize group discussions in the class. At the end of each chapter, there is a mini test including both conceptual and numerical questions.

Due to the above-mentioned features, this book is written to be used with a textbook of your choice, as a useful companion. It is particularly useful when students are preparing for their examinations. Asking and answering these conceptual questions and reviewing the key points summarized in this book is a structured approach to assess whether or not the subject has been understood and to identify the area where further revision is needed. The book is also a useful reference for those who are taking an advanced Structural and Stress Analysis course. It provides a quick recovery of the theories and important concepts that have been learnt in the past, without the need to pick up those from a more detailed and, indeed, thicker textbook.

Acknowledgements

I would like to acknowledge the constructive comments and suggestions I have received from my colleagues and students over the years of my teaching profession. These valuable suggestions have inspired me and helped me in the development of this book.

I am indebted to the external reviewers appointed by the publishers and am appreciative of their constructive criticisms. This book would not have been completed without the support received from the School of Civil Engineering at the University of Leeds, where I have been employed as an academic member of staff.

The interest of Spon Press in the publication of this book is greatly appreciated. I would like to thank Tony Moore, Matthew Gibbons, Monika Falteiskova and Katy Low for their encouragement and editorial assistance. I am also grateful for the help I received from Sunita Jayachandran and other members of staff at Integra in preparing the final version of the manuscript.

1 Introduction

Any material or structure may fail when it is loaded. The successful design of a structure requires detailed structural and stress analysis in order to assess whether or not it can safely support the required loads. Figure 1.1 shows how a structure behaves under applied loads.

To prevent structural failure, a typical design must consider the following three major aspects:

1 *Strength* – The structure must be strong enough to carry the applied loads.
2 *Stiffness* – The structure must be stiff enough such that only allowable deformation occurs.
3 *Stability* – The structure must not collapse through buckling subjected to the applied compressive loads.

The subject of structural and stress analysis provides analytical, numerical and experimental methods for determining the strength, stiffness and stability of load-carrying structural members.

1.1 Forces and moments

A *force* is a measure of its tendency to cause a body to move or translate in the direction of the force. A complete description of a force includes its *magnitude* and *direction*. The magnitude of a force acting on a structure is usually measured by Newton (N), or kilonewton (kN). In stress analysis, a force can be categorized as either external or internal. External forces include, for example, applied surface loads, force of gravity and support reactions, and the internal forces are the resisting forces generated within loaded structural elements. Typical examples of applied external forces include the following:

(a) Point load, where force is applied through a point of a structure (Figure 1.2(a))
(b) Distributed load, where force is applied over an area of a structure (Figure 1.2(b))

The *moment* of a force is a measure of its tendency to cause a body to rotate about a specific point or axis. In order to develop a moment about, for example, a specific axis, a force must act such that the body would begin to twist or bend about the axis. The magnitude of the moment of a force acting about a point or axis is directly proportional to the distance of the force from the point or axis. It is defined as the product of the force and the *lever arm*. The lever arm is the perpendicular distance between the line of action of the force and the point about which the force causes rotation. A moment is usually measured by Newton-meters (N m), or kilonewton-meters (kN m). Figure 1.3 shows how a moment about the beam–column connection is caused by the applied point load F.

Figure 1.1

Figure 1.2

Figure 1.3

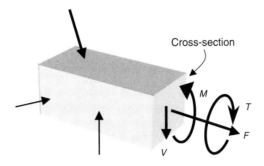

Figure 1.4

1.2   Types of force and deformation

1.2.1   Force

On a cross-section of a material subject to external loads, there exist four different types of internal force (Figure 1.4):

1   normal force, F, which is perpendicular to the cross-section;
2   shear force, V, which is parallel to the cross-section;
3   bending moment, M, which bends the material; and
4   twisting moment (torque), T, which twists the material about its central axis.

1.2.2   Deformation

Table 1.1 shows the most common types of force and their associated deformations. In a practical design, the deformation of a member can be a combination of the basic deformations shown in Table 1.1.

1.3   Equilibrium system

In static structural and stress analysis, a system in equilibrium implies that:

- the resultant of all applied forces, including support reactions, must be zero;
- the resultant of all applied moments, including bending and twisting moments, must be zero.

The two equilibrium conditions are commonly used to determine support reactions and internal forces on cross-sections of structural members.

1.3.1   The method of section

One of the most basic analyses is the investigation of the internal resistance of a structural member, that is, the development of internal forces within the member to balance the effect of the externally applied forces. The *method of section* is normally used for this purpose. Table 1.2 shows how the method of section works.

In summary, if a member as a whole is in equilibrium, any part of it must also be in equilibrium. Thus, the externally applied forces acting on one side of an arbitrary section must be balanced by the internal forces developed on the section.

Table 1.1 Basic types of deformation

Force	Deformation	Description
Normal force, Axial force, Thrust		The member is being stretched by the axial force and is in *tension*. The deformation is characterized by axial *elongation*.
Normal force, Axial force, Thrust		The member is being compressed by the axial force and is in *compression*. The deformation is characterized by axial *shortening*.
Shear force		The member is being sheared. The deformation is characterized by *distorting* a rectangle into a parallelogram.
Torque, Twist moment		The member is being twisted and is in torsion. The deformation is characterized by *angle of twist*.
Bending moment		The member is being bent and the deformation is characterized by a *bent shape*.

Table 1.2 The method of section

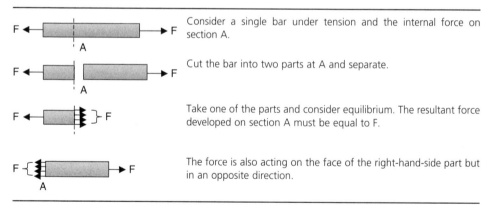

	Description
	Consider a single bar under tension and the internal force on section A.
	Cut the bar into two parts at A and separate.
	Take one of the parts and consider equilibrium. The resultant force developed on section A must be equal to F.
	The force is also acting on the face of the right-hand-side part but in an opposite direction.

1.3.2 The method of joint

The analysis or design of a truss requires the calculation of the forces in each of its members. Taking the entire truss as a free body, the forces in the members are internal forces. In order to determine the internal forces in the members jointed at a particular joint, for example, joint A in Figure 1.5, the joint can be separated from the truss system by cutting all the members around it. On the sections of the cuts there exist axial forces that can be further determined by considering the equilibrium of the joint under the action of the internal forces and the externally applied loads at the joint, that is, by resolving the forces in the x and y directions, respectively, and letting the resultants be zero.

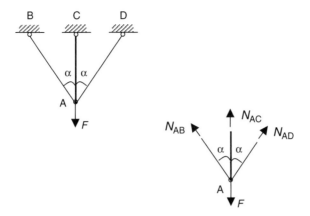

Figure 1.5

1.4 Stresses

Stress can be defined as the intensity of internal force that represents internal force per unit area at a point on a cross-section. Stresses are usually different from point to point. There are two types of stresses, namely *normal* and *shear stresses*.

1.4.1 Normal stress

Normal stress is a stress perpendicular to a cross-section or cut. For example, for the simple structural element shown in Figure 1.6(a), the normal stress on section m–m can be calculated as

$$\text{Normal stress } (\sigma) = \frac{\text{force (on section m–m)}}{\text{area (of section m–m)}} \qquad (1.1a)$$

The basic unit of stress is N/m², which is also called a Pascal.

  In general a stress varies from point to point (Figure 1.6(b)). A general stress can be calculated by

Figure 1.6(a)

Figure 1.6(b)

$$stress\ at\ point\ P = \lim_{\Delta A \to 0} \frac{\Delta F}{\Delta A} \qquad\qquad (1.1b)$$

where ΔF is the force acting on the infinitesimal area, ΔA, surrounding P.

1.4.2 Shear stress

Shear stress is a stress parallel to a cross-section or cut. For example, for the plates connected by a bolt, shown in Figure 1.7(a), the forces are transmitted from one part of structure to the other part by causing stresses in the plane parallel to the applied forces. These stresses are shear stresses. To compute the shear stresses, a cut is taken through the parallel plane and uniform distribution of the stresses over the cut is assumed. Thus:

$$\tau = \frac{force}{area} = \frac{P}{A} \qquad\qquad (1.2)$$

where A is the cross-sectional area of the bolt.

At a point in a material, shear stresses always appear in pair acting on two mutually perpendicular planes. They are equal in magnitude, but in an opposite sense, that is, either towards or away from the point (Figure 1.7(b)).

From the definition of normal and shear stresses, the following three characteristics must be specified in order to define a stress:

1   the magnitude of the stress;
2   the direction of the stress; and
3   the plane (cross-section) on which the stress is acting.

Figure 1.7(a)

Figure 1.7(b)

Figure 1.8

1.5   Strains

Strain is a measure of relative deformation. Strains can be categorized as *normal* and *shear* strains.

   Normal strain is a measure of the change in length per unit length under stress (Figure 1.8). It is measured by the following formula:

$$\text{Normal strain}(\varepsilon) = \frac{\text{change in length}}{\text{original length}} = \frac{\Delta L}{L} \tag{1.3}$$

   Shear strain is a measure of the change caused by shear stresses in the right angle between two fibres within a plane (Figure 1.9).

$$\text{Shear strain} \quad \gamma = \alpha_1 + \alpha_2 \tag{1.4}$$

Shear strains are dimensionless.

1.6   Strain–stress relation

Strains and stresses always appear in pair and their relationship depends on the properties of materials. Strain–stress relation is also termed as *Hooke's law*, which determines how much strain occurs under a given stress. For materials undergoing linear elastic deformation, stresses are proportional to strains. Thus, for the simple load cases shown in Figures 1.8 and 1.9, the strain–stress relations are:

$$\sigma = E\varepsilon$$
$$\tau = G\gamma \tag{1.5}$$

Figure 1.9

(a) bar before loading

(b) bar after loading

Figure 1.10

where E is called *modulus of elasticity* or *Young's modulus*. G is termed as *shear modulus*. They are all material-dependent constants and measure the unit of stress, for example, N/mm², since strains are dimensionless. For isotropic materials, for example, most metals, E and G have the following relationship:

$$G = \frac{E}{2(1+v)} \tag{1.6}$$

In Equation (1.6), v is called *Poisson's ratio*, which is also an important material constant. Figure 1.10 shows how a Poisson's ratio is defined by comparing axial elongation and lateral contraction of a prismatic bar in tension. Poisson's ratio is defined as:

$$\begin{aligned} \text{Poisson's ratio } (v) &= \left| \frac{\text{lateral strain}}{\text{axial strain}} \right| \\ &= -\frac{\text{lateral strain (contraction)}}{\text{axial strain (tension)}} \end{aligned} \tag{1.7}$$

A negative sign is usually assigned to a contraction. Poisson's ratio is a dimensionless quality that is constant in the elastic range for most materials and has a value between 0 and 0.5.

1.7   Generalized Hooke's law

Generalized Hooke's law is an extension of the simple stress–strain relations of Equation (1.5) to a general case where stresses and strains are three-dimensional.

Consider a cube subjected to normal stresses, σ_x, σ_y and σ_z, in the directions of x, y, and z coordinate axes, respectively (Figure 1.11(a)).

From Figure 1.11, we have

Strain of Case (a) = strain of Case (b) + strain of Case (c) + strain of Case (d)

In particular, considering the normal strain of Case (a) in the x direction and applying Equation (1.5) and Equation (1.7) to Cases (b), (c) and (d), we have

Normal strain in the x direction		
By σ_x Figure 1.11(b)	By σ_y Figure 1.11(c)	By σ_z Figure 1.11(d)
$\varepsilon_x = \frac{\sigma_x}{E}$	$v = -\frac{\varepsilon_x}{\varepsilon_y}$ $\varepsilon_x = -v\varepsilon_y$	$v = -\frac{\varepsilon_x}{\varepsilon_z}$ $\varepsilon_x = -v\varepsilon_z$

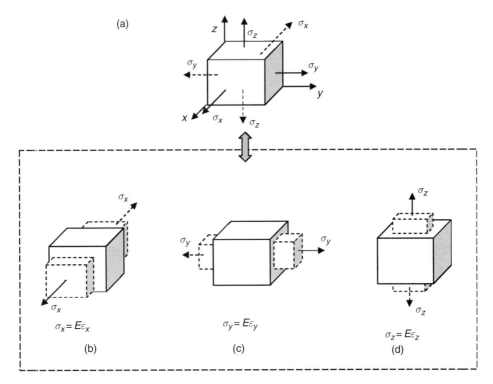

Figure 1.11

Thus, the normal strain of Case (a) in the x direction is as follows:

$$\varepsilon_x = \frac{\sigma_x}{E} - \nu\varepsilon_y - \nu\varepsilon_z$$

From Figures 1.11(c) and 1.11(d):

$$\varepsilon_y = \frac{\sigma_y}{E},$$

$$\varepsilon_z = \frac{\sigma_z}{E}$$

Then:

$$\varepsilon_x = \frac{\sigma_x}{E} - \nu\frac{\sigma_y}{E} - \nu\frac{\sigma_z}{E} = \frac{1}{E}\left[\sigma_x - \nu(\sigma_y + \sigma_z)\right] \qquad (1.8a)$$

The strains in the y and z directions can also be calculated by following exactly the same procedure described above:

$$\varepsilon_y = \frac{\sigma_y}{E} - \nu\frac{\sigma_x}{E} - \nu\frac{\sigma_z}{E} = \frac{1}{E}\left[\sigma_y - \nu(\sigma_x + \sigma_z)\right]$$

$$\qquad (1.8b)$$

$$\varepsilon_z = \frac{\sigma_z}{E} - \nu\frac{\sigma_x}{E} - \nu\frac{\sigma_y}{E} = \frac{1}{E}\left[\sigma_z - \nu(\sigma_x + \sigma_y)\right]$$

For a three-dimensional case, shear stresses and shear strains may occur within three independent planes, that is, in the x–y, x–z and y–z planes, for which the following three shear stress and strain relations exist:

$$\gamma_{xy} = \frac{\tau_{xy}}{G}$$

$$\gamma_{xz} = \frac{\tau_{xz}}{G} \tag{1.8c}$$

$$\gamma_{yz} = \frac{\tau_{yz}}{G}$$

Equation (1.8) is the generalized Hooke's law. The application of Equation (1.8) is limited to isotropic materials in the linear elastic range.

The generalized Hooke's law of Equation (1.8) represents strains in terms of stresses. The following equivalent form of Hooke's law represents stresses in terms of strains:

$$\sigma_x = \frac{E(1-v)}{(1+v)(1-2v)} \left[\varepsilon_x + \frac{v}{1-v}(\varepsilon_y + \varepsilon_z) \right]$$

$$\sigma_y = \frac{E(1-v)}{(1+v)(1-2v)} \left[\varepsilon_y + \frac{v}{1-v}(\varepsilon_x + \varepsilon_z) \right]$$

$$\sigma_z = \frac{E(1-v)}{(1+v)(1-2v)} \left[\varepsilon_z + \frac{v}{1-v}(\varepsilon_y + \varepsilon_x) \right] \tag{1.9}$$

$$\tau_{xy} = G\gamma_{xy}$$

$$\tau_{xz} = G\gamma_{xz}$$

$$\tau_{yz} = G\gamma_{yz}$$

1.8   Strength, stiffness and failure

Failure is a condition that prevents a material or a structure from performing the intended task. For the cantilever shown in Figure 1.12, the following two questions, for example, can be asked:

(i)   What is the upper limit of stress that can be reached in the material of the beam?

The answer to this question provides a strength criterion that can be adopted in the design of the beam (Figure 1.13):

- An upper limit at which the stress–strain relationship departs from linear is called the *proportional limit, σ_{pl}*.
- An upper limit at which permanent deformation starts is called the *yield strength, σ_{Yield}*.
- An upper limit, that is the maximum stress a material can withstand is called the *ultimate strength, σ_u*.

Strength is a property of material.

(ii)   What is the maximum tip deflection that is acceptable?

The answer to this question provides a stiffness design criterion that represents the *stiffness* or the resistance of an elastic body to deformation.

Factors that influence stiffness of a structural member include material modulus, structural configuration and mode of loading. For example, the tip deflection of a cantilever varies if the materials, length, shape of cross-section or the applied load change.

Figure 1.12

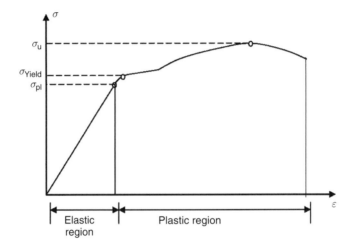

Figure 1.13

Strength and stiffness are measurements of resistance to failure. Violation of any of the above criterions is defined as failure. In a typical design, a primary task is to choose materials and member dimensions such that:

- stresses are maintained below the limits for the chosen materials;
- deformations are maintained below the limits for the structure application.

1.9   Key points review

- An applied force can be in the form of point load, distributed load or moment.
- An applied load causes deformation and eventually failure of a structure.
- An applied force causes internal forces/stresses.
- Stress is defined as intensity of internal force at a point of material.
- A stress has magnitude and direction, and is always related to a special plane (cross-section).
- Normal stress is a stress that is perpendicular to a cross-section and causes tension or compression.

- Shear stress is a stress that is parallel to a cross-section and causes distortion or twisting.
- Strain is a measurement of relative deformation at a point of material, and is a nondimensional quantity.
- Normal strain represents either an elongation or a contraction.
- Shear strain is measurement of distortion, measured by change of a right angle.
- The relationship between stresses and strains depends on properties of materials. For linear elastic materials, the relationship is called Hooke's law.
- For a linearly elastic and isotropic material, E, G and ν are related and only two of them are independent.
- Different materials normally have different strength and the strength depends only on property of material.
- Stiffness of a member depends on not only property of material, but also geometrical and loading conditions; stiffness is not a property of material.
- Proportional limit, σ_{pl}, is the upper limit at which the stress–strain relationship departs from linear.
- Yield strength, σ_{Yield}, is the upper limit at which permanent deformation starts.
- Ultimate strength, σ_u, is the maximum stress a material can withstand.

1.10 Basic approach for structural analysis

The solution of a stress problem always follows a similar procedure that is applicable for almost all types of structures. Figure 1.14 presents a flow chart for the procedure. In general, either deformations (strains) or forces (stresses) are the quantities that need to be computed in a structural analysis of design. The following steps represent a general approach to the solution of a structural problem:

- Calculating support reactions is usually a start point of a stress analysis. For a statically determinate structure, the reactions are determined by the application of equilibrium equations. For a statically indeterminate structure, additional equations must be sought.
- If forces or stresses on a section are wanted, the method of section or/and the method of joint are used to cut through the section so that the structure is cut into two parts, i.e., two free bodies.
- A part on either side of the section is taken as a free body and required to satisfy the equilibrium conditions. On the section concerned, the internal forces that keep the part in equilibrium include, in a general case, an normal force, a shear force, a bending moment and a twist moment (torque). These internal forces are found by the application of static equilibrium of all forces acting on the free body.
- Once the internal forces on the section are determined, the stresses caused by the forces can be calculated using appropriate formulas of stress analysis.
- From the stress solutions, Hooke's law can be used to compute strains and then displacements, e.g., deflection of a beam subjected to bending.
- Both the stress and the strain solutions are further used in design to meet relevant strength and stiffness criterions.

Figure 1.14

1.11 Conceptual questions

1. What is the difference between applied loads and reactions?
2. What is meant by 'stress' and why is it a local measurement of force?
3. What is the unit for measuring stress?
4. What is the difference between a normal stress and a shear stress?
5. What is meant by 'strain'? Is it a local measure of deformation?
6. Is a 'larger strain' always related to a 'larger displacement'?
7. What is the physical meaning of 'shear strain'?
8. What is the unit for measuring strain?
9. What is Young's modulus and how can it be determined from a simple tension test?
10. What is the simplest form of stress and strain relationship?
11. If the displacement at a point in a material is zero, the strain at the same point must be zero. Is this correct and why?
12. How is Poisson's ratio defined?
13. For a linearly elastic and isotropic material, what is the relationship between Young's modulus, shear modulus and Poisson's ratio?
14. How can 'failure' of a structural member be defined?
15. What are meant, respectively, by 'proportional limit', 'yield strength' and 'ultimate strength'? Are they properties of material?
16. What is meant by 'stiffness'? Is it a property of materials and why?
17. Describe how the method of joint can be used in structural analysis.

18. Describe how the method of section can be used in structural analysis.
19. The stress–strain curve for a hypothetical material is given below. If the strain at the top and bottom of a section are, respectively, $2\varepsilon_m$ in tension and $-2\varepsilon_m$ in compression, sketch the stress distribution over the height of the section.

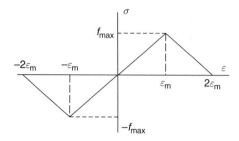

Figure Q1.19

1.12   Mini test

Problem 1.1:   Which one of the following statements is correct?

A. Normal stress on a cross-section is always equal to the internal force acting on the section divided by the cross-sectional area.

B. Normal stress on a cross-section is always NOT equal to the force acting on the section divided by the cross-sectional area.

C. The internal normal force acting on a cross-section is the resultant of the normal stresses acting on the same section.

D. Normal stresses on a cross-section are independent of the normal force acting on the same section.

Problem 1.2:   What are the differences between displacement, deformation and strain? And which one of the following statements is correct in relation to the loaded beam shown in the figure?

Figure P1.2

A. There are displacement, deformation and strain in BC.

B. There is only displacement in BC.

C. There are both displacement and deformation but without strain in BC.

D. There are no displacement, deformation and strain in BC.

Problem 1.3:   Are the longitudinal normal stress, strain and internal normal force on the cross-sections of the bar (Figure P1.3) constant along its axis? Can the strain of the bar be calculated by $\varepsilon = \Delta L/L$ and why?

If the bar has a circular section whose largest and smallest diameters are D and d, respectively, calculate the strain along the bar. Assume that the Young's modulus of the materials is E.

Figure P1.3

Problem 1.4:   Use the method of section to determine the internal forces on the cross-section at B (Figure P1.4).

Figure P1.4

Problem 1.5:   Use the method of joint to determine the axial forces in the members of the truss shown in the figure.

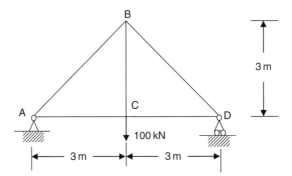

Figure P1.5

2 Axial tension and compression

In practical situations, axial tension or compression is probably the simplest form of deformation. This type of deformation is characterized by the following:

- The action line of the resultant of applied forces coincides with the axis of the member.
- Under the axial force, normal stress develops on cross-sections.
- Under the axial force, the deformation of the member is dominated by either axial elongation or axial shortening with associated contraction or expansion, respectively, in the lateral direction.

2.1 Sign convention

A positive axial force (stress) is defined as a force (stress) that induces elongation (Figure 2.1(a)). A negative axial force (stress) is defined as a force (stress) that induces axial shortening (Figure 2.1(b)).

   The sign convention is designed to characterize the nature of the force or stress, rather than in relation to a particular direction of the coordinates. For example, in Figure 2.1(a), both forces are positive because they are all tensile. While setting up equilibrium equation, the two forces are opposite, that is, one is positive and the other is negative.

2.2 Normal (direct) stress

The uniformly distributed normal stress, σ, on section m is calculated by:

$$\sigma = \frac{F}{A} \tag{2.1a}$$

where A is the cross-sectional area of the bar; σ takes the sign of F (Figure 2.2). Since the force and the cross-sectional area are both constant along the bar in this case, the normal stress is also constant along the bar. This is not always true if either the force or the cross-sectional area is variable.

   Applying the simple form of Hooke's law (Equation (1.5)) to the bars yields:

$$\varepsilon = \frac{\sigma}{E} = \frac{F}{EA} \tag{2.1b}$$

Figure 2.1

Figure 2.2

Again this applies only when both the internal normal force and the cross-sectional area are constant along the bar.

2.3   Stresses on an arbitrarily inclined plane

There are situations where stresses on a plane that is not perpendicular to the member axis are of interest, for example, the direct stress σ_α and shear stress τ_α along the interface of the adhesively bonded scarf joint shown in Figure 2.3.
     From the equilibrium of:

$$\sigma = \frac{F}{A}$$

Normal stress (peeling stress) $\sigma_\alpha = \dfrac{F}{A}\cos^2\alpha = \sigma\cos^2\alpha$ (2.2)

Shear stress $\tau_\alpha = \dfrac{F}{2A}\sin 2\alpha = \dfrac{\sigma}{2}\sin 2\alpha$ (2.3)

Resultant stress $p_a = \dfrac{F}{A}\cos\alpha = \sigma\cos\alpha$ (2.4)

Important observations:

(a)   The maximum normal stress occurs when $\alpha = 0$, i.e., $\sigma_{max} = \sigma$.
(b)   The maximum shear stress occurs when $\alpha = \pm 45°$, i.e., $\tau_{max} = \sigma/2$.
(c)   On the cross-section where maximum normal stress occurs ($\alpha = 0$), there is no shear stress.

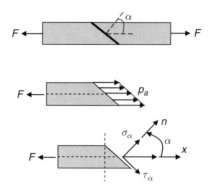

Figure 2.3

2.4   Deformation of axially loaded members

2.4.1   Members of uniform sections

Equations (1.1), (1.5) and (2.1) are sufficient to determine the deformation of an axially loaded member.

Figure 2.4

Axial deformation:

$$\Delta l = l_1 - l = \frac{Nl}{EA} \tag{2.5}$$

where N is the internal axial force due to the action of F ($N = F$ in Figure 2.4).

When the internal axial force or/and the cross-sectional area vary along the axial direction:

$$\Delta l = \int_l \frac{N(x)}{EA(x)} dx \tag{2.6}$$

2.4.2   Members with step changes

An axially loaded bar may be composed of segments with different cross-sectional areas, internal normal forces and even materials, as shown in Figure 2.5.

The internal axial forces, and therefore the normal stress on the cross-sections, will not be constant along the bar (except Figure 2.5(c), where normal strains are different). They are constant within each segments, in which cross-sectional areas, internal normal stresses and materials are all constant. Thus, to determine the stress, strain and elongation of such a bar, each segment must be considered independently.

Figure 2.5

2.5 Statically indeterminate axial deformation

There are axially loaded structural members whose internal forces cannot be simply determined by equations of equilibrium. This type of structure is called *statically indeterminate structures*. For the bar shown in Figure 2.6, for example, the two reaction forces at the fixed ends cannot be uniquely determined by considering only the equilibrium in the horizontal direction. The equilibrium condition gives only a single equation in terms of the two unknown reaction forces, R_A and R_B. An additional condition must be sought in order to form the second equation. Usually, for a statically indeterminate structure, additional equations come from considering deformation of the system.

In Figure 2.6, one can easily see that the overall elongation of the bar is zero since it is fixed at both ends. This is called *geometric compatibility of deformation*, which provides an additional equation for the solution of this problem. The total deformation of the bar is calculated considering the combined action of the two unknown support reactions and the externally applied loads. Example 2.3 shows how this condition can be used to form an equation in terms of the two reaction forces. In general, the following procedure can be adapted:

* Replace supports by reaction forces.
* Establish static equilibrium equation.
* Consider structural deformation, including deformation of members, to form a geometric relationship (equation).
* Use the force-deformation relationships, e.g., Equation (2.5) for axial deformation, and introduce them to the geometric equation.
* Solve a simultaneous equation system that consists of both the static equilibrium equations and the geometric compatibility equations.

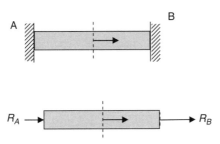

Figure 2.6

2.6   Elastic strain energy of an axially loaded member

Strain energy is the internal work done in a body by externally applied forces. It is also called *the internal elastic energy of deformation.*

2.6.1   Strain energy U in an axially loaded member

For an axially loaded member with constant internal axial force:

$$U = \frac{N\Delta l}{2} = \frac{N^2 l}{2EA} \qquad (2.7a)$$

If the member is composed of segments with different cross-sectional areas, A_i, internal normal forces, N_i, and materials, E_i, the strain energy is the sum of the energy stored in each of the segments as:

$$U = \sum_{i=1}^{n} \frac{N_i \Delta l_i}{2} = \sum_{i=1}^{n} \frac{N_i^2 l_i}{2E_i A_i} \qquad (2.7b)$$

2.6.2   Strain energy density, U_0

Strain energy density is the strain energy per unit volume. For the axially loaded member with constant stress and strain:

$$U_0 = \frac{U}{volume} = \frac{N^2 l/2EA}{Al}$$

$$= \frac{\sigma \varepsilon}{2} = \frac{E\varepsilon^2}{2} = \frac{\sigma^2}{2E} \qquad (2.8)$$

U_0 is usually measured in J/m^3.

2.7   Saint-Venant's principle and stress concentration

Equation (2.1) assumes that normal stress is constant across the cross-section of a bar under an axial load. However, the application of this formula has certain limitations.

When an axial load is applied to a bar of a uniform cross-section as shown in Figure 2.7(a), the normal stress on the sections away from the localities of the applied concentrated loads will be uniformly distributed over the cross-section. The normal stresses on the sections near the two ends, however, will not be uniformly distributed because of the nonuniform deformation caused by the applied concentrated loads. Obviously, a higher level of strain, and hence, higher level of stress, will be induced in the vicinity of the applied loads.

Figure 2.7(a)

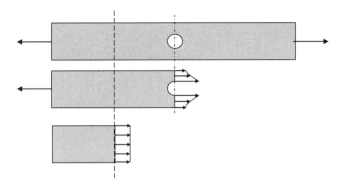

Figure 2.7(b)

This observation is often useful when solving static equilibrium problems. It suggests that if the distribution of an external load is altered to a new distribution that is however *statically equivalent* to the original one, that is with the same resultant forces and moments, the stress distribution on a section sufficiently far from where the alteration was made will be little affected. This conclusion is termed as *Saint-Venant's principle*.

The unevenly distributed stress in a bar can also be observed if a hole is drilled in a material (Figure 2.7(b)). The stress distribution is not uniform on the cross-sections near the hole. Since the material that has been removed from the hole is no longer available to carry any load, the load must be redistributed over the remaining material. It is not redistributed evenly over the entire remaining cross-sectional area, with a higher level of stress near the hole. On the cross-sections away from the hole, the normal stress distribution is still uniform. The unevenly distributed stress is called *stress concentration*.

This observation suggests that if a structural member has a sudden change of cross-section shape or discontinuity of geometry, stress concentration will occur in the vicinity of the sudden changes or discontinuities.

2.8   Stresses caused by temperature

In a statically determinate structure, the deformation due to temperature changes is usually disregarded, since in such a structure the members are free to expand or contract. However, in a statically indeterminate structure, expansion or contraction can be restricted. This sometimes can generate significant stresses that may cause failure of a member and eventually the entire structure.

For a bar of length L (Figure 2.8), the free deformation caused by a change in temperature, ΔT, is:

$$\Delta L_T = \alpha_T \Delta T L \tag{2.9a}$$

and the thermal strain is, therefore:

$$\varepsilon_T = \alpha_T \Delta T \tag{2.9b}$$

where α_T is the coefficient of thermal expansion.

Figure 2.8

If the bar is not completely free to expand or contract, the strain of the bar is a sum of the above thermal strain and the strain caused by the stress that is developed due to the restraints to free expansion, that is:

$$\varepsilon = \frac{\sigma}{E} + \alpha_T \Delta T \qquad\qquad (2.10)$$

2.9 Key points review

- A shaft/rod is long compared to its other two dimensions.
- The load applied is along the axial direction.
- If a shaft is subjected to axially applied loads, the deformation can be defined by either axial tension or axial compression.
- The cross-section deforms uniformly.
- If a structural member is free to expand and contract, a change in temperature will generate strains, but not stresses.
- If a structural member is prevented to expand and contract, a change in temperature will generate both strains and stresses.
- The internal stresses and forces on sections perpendicular to the axis are normal stresses and axial forces, respectively.
- For an axially loaded bar, the maximum normal stress at a point is the normal stress on the section perpendicular to the axis.
- For an axially loaded bar, the maximum shear stress at a point is the shear stress acting on the plane that is 45° to the axis.
- On the cross-section where maximum normal stress occurs, there exists no shear stress.
- On the plane of maximum shear stress, the normal stress is not necessarily zero.
- The resultant of normal stress on a cross-section is the axial force acting on the same section.
- The axial tension or compression stiffness of a section is *EA*.
- Strain energy is proportional to square of forces, stresses or strains.
- For a structure system, if the number of independent equilibrium equations is less than the number of unknown forces, the system is termed as an *indeterminate system*.
- To solve a statically indeterminate system, geometrical compatibility of deformation must be considered along with the equilibrium conditions.
- Replacing a load applied on a material by an alternative, but *statically equivalent* load will affect only the stress field in the vicinity.
- An abrupt change of geometry of a structural member will cause stress concentration near the region of the change.

2.10   Recommended procedure of solution

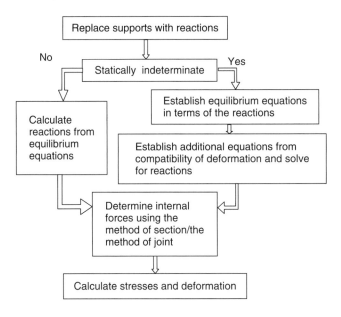

2.11   Examples

EXAMPLE 2.1

The uniform bar is loaded as shown in Figure E2.1(a). Determine the axial stress along the bar and the total change in length. The cross-sectional area and the Young's modulus of the bar are, respectively, $A = 1\,cm^2$ and $E = 140\,GPa$.

Figure E2.1(a)

[Solution] *This is a uniform bar with forces applied between the two ends. The internal forces along the bar are, therefore, not constant. The stresses and elongation between A and B, B and C, and C and D must be calculated individually and the total change in length is equal to the sum of all the elongations. The method of section is used to determine the internal forces.*

Between A and B:

Figure E2.1(b)

From the equilibrium of Figure E2.1(b):

$$N_{AB} - 5\,kN = 0$$

$$N_{AB} = 5\,kN$$

The normal stress between A and B (Equation (2.1)):

$$\sigma_{AB} = \frac{N_{AB}}{A} = \frac{5\,kN}{0.0001\,m^2} = 5 \times 10^4\,kN/m^2 = 50\,MPa$$

The elongation between A and B (Equation (2.5)):

$$\Delta_{AB} = \frac{N_{AB}L_{AB}}{EA} = \frac{5\,kN \times 0.5\,m}{140 \times 10^6\,kN/m^2 \times 0.0001\,m^2} = 178.57 \times 10^{-6}\,m$$

Between B and C:

Figure E2.1(c)

From the equilibrium of Figure E2.1(c):

$$N_{AB} + 8\,kN - 5\,kN = 0$$

$$N_{AB} = -8\,kN + 5\,kN = -3\,kN\ (\text{compression})$$

The normal force should be in the opposite direction of the assumed N_{BC}.
The normal stress between B and C (Equation (2.1)):

$$\sigma_{BC} = \frac{N_{BC}}{A} = \frac{-3\,kN}{0.0001\,m^2} = -3 \times 10^4\,kN/m^2 = -30\,MPa$$

The elongation between B and C (Equation (2.5)):

$$\Delta_{BC} = \frac{N_{BC}L_{BC}}{EA} = \frac{-3\,kN \times 0.75\,m}{140 \times 10^6\,kN/m^2 \times 0.0001\,m^2} = -160.71 \times 10^{-6}\,m$$

Between C and D:

Figure E2.1(d)

From the equilibrium of Figure E2.1(d):

$$N_{CD} + 8\,kN + 4\,kN - 5\,kN = 0$$

$$N_{AB} = -8\,kN - 4\,kN + 5\,kN = -7\,kN \text{ (compression)}$$

The normal force should be in the opposite direction of the assumed N_{CD}.
 The normal stress between C and D (Equation (2.1)):

$$\sigma_{CD} = \frac{N_{CD}}{A} = \frac{-7\,kN}{0.0001\,m^2} = -7 \times 10^4\,kN/m^2 = -70\,MPa$$

The elongation between C and D (Equation (2.5)):

$$\Delta_{CD} = \frac{N_{CD}L_{CD}}{EA} = \frac{-7\,kN \times 1.0\,m}{140 \times 10^6\,kN/m^2 \times 0.0001\,m^2} = -500 \times 10^{-6}\,m$$

The total change in length of the bar is:

$$\Delta_{Total} = \Delta_{AB} + \Delta_{BC} + \Delta_{CD} = 178.57 \times 10^{-6}\,m - 160.71 \times 10^{-6}\,m - 500 \times 10^{-6}\,m$$

$$= -482.14 \times 10^{-6}\,m = -0.482\,mm$$

The negative sign means the change in length is contraction.

EXAMPLE 2.2
The concrete pier shown in Figure E2.2 supports a uniform pressure of $20\,kN/m^2$ at the top. The density of the concrete is $25\,kN/m^3$ and the pier is 0.5 m thick. Calculate the reaction force at the base and the stress at a level of 1 m above the base.

Figure E2.2

[Solution] *This is a statically determinate structure subjected to only vertical loads. The vertical reaction at the base can be determined from vertical equilibrium. The method of section can be used to calculate the stress on section m–m.*

From the equilibrium in the vertical direction, the reaction R must be equal to the sum of the weight and the resultant of the pressure.

Weight of the pier:

$$W = \frac{(0.5+1.5) \times 2}{2} \times 0.5 \times 25 = 25\,kN$$

Resultant of the pressure:

$$P = 0.5 \times 0.5 \times 20 = 5\,kN$$

Thus, the reaction at base:

$$R = W + P = 30\,kN$$

Cut at m–m and take the upper part of the pier as a free body (Figure E2.2(b)). Forces from the part above the cut:

$$P + W = 5 + \frac{(0.5+1) \times 1}{2} \times 0.5 \times 25 = 14.4\,kN$$

From the vertical equilibrium of Figure E2.2(b):

$$N = P + W = 14.4\,kN$$

Thus, the stress at this level is:

$$\sigma = \frac{N}{A} = \frac{14.4\,kN}{1.5\,m \times 0.5\,m} = 19.2\,kN/m^2 \text{ (compression)}$$

EXAMPLE 2.3
A bar made of steel and brass has the dimensions shown in Figure E2.3. The bar is rigidly fixed at both ends. Calculate the end reactions and stresses when the force F is applied at level C. Let $A_{steel} = 2 \times 10^4\,mm^2$, $E_{steel} = 200\,GPa$, $A_{br} = 1 \times 10^4\,mm^2$, $E_{br} = 100\,GPa$ and $F = 1 \times 10^3\,kN$.

Figure E2.3(a)

[Solution] *This is a statically indeterminate system subjected to only vertical forces. The equilibrium in the vertical direction must be first established. Due to the two fixed ends, the total elongation of the bar is zero. This observation provides the additional equation in terms of the two unknown reactions.*

Figure E2.3(b)

From the vertical equilibrium of Figure E2.3(b):

$$R_1 + R_2 - F = 0 \qquad\qquad\qquad\qquad \text{(E2.3a)}$$

Since the total deformation between A and D is zero:

$$\Delta l_{AB} + \Delta l_{BC} - \Delta l_{CD} = 0$$

where Δl_{AB}, Δl_{BC} and Δl_{CD} are the respective deformations between A and B, B and C, and C and D. The negative sign before Δl_{CD} denotes compressive deformation between C and D. From Equation (2.5):

$$\Delta l_{AB} = \frac{R_2 \times 2a}{E_{steel} A_{steel}},$$

$$\Delta l_{BC} = \frac{R_2 a}{E_{br} A_{br}},$$

$$\Delta l_{CD} = \frac{R_1 a}{E_{br} A_{br}}$$

Thus:

$$\frac{R_2 \times 2a}{E_{steel} A_{steel}} + \frac{R_2 a}{E_{br} A_{br}} - \frac{R_1 a}{E_{br} A_{br}} = 0$$

or

$$\frac{2R_2}{200 \times 10^9 \, \text{Pa} \times 100 \times 10^{-4} \, \text{m}^2} + \frac{R_2 - R_1}{100 \times 10^9 \, \text{Pa} \times 200 \times 10^{-4} \, \text{m}^2} = 0$$

or

$$3R_2 - R_1 = 0 \qquad\qquad\qquad\qquad \text{(E2.3b)}$$

From Equations (E2.3a) and (E2.3b):

$$R_1 = \frac{3}{4}F = 750 \, \text{kN} \quad R_2 = \frac{1}{4}F = 250 \, \text{kN}$$

Stresses in the bar:

$$\text{Stress between A and B } \sigma_{AB} = \frac{R_2}{A_{steel}} = \frac{250 \, \text{kN}}{100 \times 10^{-4} \, \text{m}^2} = 25 \, \text{MPa (tension)}$$

Stress between B and C $\sigma_{BC} = \dfrac{R_2}{A_{br}} = \dfrac{250\,kN}{200 \times 10^{-4}\,m^2} = 12.5\,MPa$ (tension)

Stress between C and D $\sigma_{CD} = \dfrac{R_1}{A_{br}} = \dfrac{750\,kN}{200 \times 10^{-4}\,m^2} = 37.5\,MPa$ (compression)

EXAMPLE 2.4

A rigid beam of 3a long is hinged at one end and supported by two steel wires as shown in Figure E2.4. Wire 1 is 0.1 mm short due to a manufacturing error and has to be stretched so as to be connected to the beam. If the ratio between the cross-sectional areas of the two wires, that is, and A_1/A_2, is 2 and the allowable stress of steel is 160 MPa, calculate the minimum cross-sectional areas of both wires. $E_{steel} = 200\,GPa$.

Figure E2.4

[Solution] *This is a statically indeterminate structure. Apart from the equilibrium conditions, the relationship between the elongation of wires 1 and 2 can be used as the compatibility condition. Note that the beam can only rigidly rotate about the pin.*

From the equilibrium of moment about the pinned end (Figure E2.4(b)):

$$F_1 a + 2F_2 a - 12 \times 3a = 0 \qquad\qquad (E2.4a)$$

From the deformation shown in Figure E2.4(c):

$$\Delta l_2 = 2(\Delta l_1 - 0.1)$$

where Δl_1 and Δl_2 represent, respectively, the elongation of wires 1 and 2. Thus, from Equation (2.5):

$$\Delta l_1 = \dfrac{F_1 l_1}{E_{steel} A_1}$$

$$\Delta l_2 = \frac{F_2 l_2}{E_{steel} A_2}$$

then

$$\frac{F_2 l_2}{E_{steel} A_2} = 2\left(\frac{F_1 l_1}{E_{steel} A_1} - 0.1\right)$$

or

$$\frac{F_2 \times 1 \times 10^3}{200 A_2} = 2\left(\frac{F_1 \times 999}{2 \times 200 \times A_2} - 0.1\right) \tag{E2.4b}$$

In Equation (E2.4b) A_1 is replaced by $2A_2$, and kN and mm are used as the basic units. Solving Equations (E2.4a) and (E2.4b) simultaneously yields:

$$F_1 = 12 + 0.0267 A_2$$

$$F_2 = 12 - 0.0267 A_2$$

To meet the design requirement:

$$\frac{F_1}{A_1} = \frac{12 + 0.0267 A_2}{A_1} = \frac{12 + 0.0267 A_2}{2A_2} \leq 160 \times 10^{-3}\, \text{kN/mm}^2 \tag{E2.4c}$$

$$\frac{F_2}{A_2} = \frac{12 - 0.0267 A_2}{A_2} = \frac{12 - 0.0133 A_2}{A_2} \leq 160 \times 10^{-3}\, \text{kN/mm}^2 \tag{E2.4d}$$

From Equation (E2.4c):

$$A_2 = 40.91\, \text{mm}^2$$

From Equation (E2.4d):

$$A_2 = 69.24\, \text{mm}^2$$

To meet both requirements, $A_2 = 69.24\, \text{mm}^2$ and $A_1 = 2A_2 = 138.48\, \text{mm}^2$ are the respective minimum areas of the supporting wires.

EXAMPLE 2.5

A concrete cylinder of area A_c is reinforced concentrically with a steel bar of area A_{steel}. The composite unit is L long and subjected to a uniform temperature change of ΔT. Compute the thermal stresses in the concrete and steel. The thermal coefficients and Young's modulus of the concrete and steel are, respectively, α_c and α_{steel} and E_c and E_{steel}.

Figure E2.5

[Solution] *Due to the difference in the coefficients of thermal expansion, the incompatibility of the longitudinal thermal deformation between concrete and steel will cause internal stresses in both concrete and steel. When the concrete and steel work as a unit, the final expansion of the cylinder, indicated by l_{final} in Figure E2.5(a), is the result of both free thermal expansion, Δl_c^{free} and Δl_{steel}^{free}, and the deformation due to the stresses induced by the incompatibility Δ_{N_c} and $\Delta_{N_{steel}}$. For both concrete and steel, the relationships between these deformations need to be established for the solution. Because this is a symmetric system, only half of the cylinder is considered.*

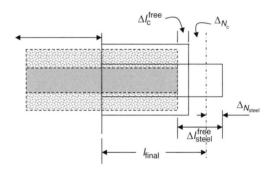

Figure E2.5(a)

(a) *Thermal expansion of the concrete and steel if they are not bonded together*
     For the concrete, from Equation (2.9):

$$\Delta l_c^{free} = \alpha_c \Delta T \frac{L}{2}$$

   For the steel:

$$\Delta l_{steel}^{free} = \alpha_{steel} \Delta T \frac{L}{2}$$

(b) *Deformation due to thermal stresses*
     Assume that the stresses on the cross-sections of both concrete and steel are uniform. The resultants of these stresses in concrete and steel are, respectively, N_c and N_{steel}. Thus the deformation due to these forces are as follows:
     For the concrete:

$$\Delta_{N_c} = \frac{N_c L/2}{E_c A_c}$$

   For the steel:

$$\Delta_{N_{steel}} = \frac{N_{steel} L/2}{E_{steel} A_{steel}}$$

(c) *Geometric relation between the above deformations*
From Figure E2.5(a):

$$\Delta_{N_c} + \Delta_{N_{steel}} = \Delta l^{free}_{steel} - \Delta l^{free}_c$$

or

$$\frac{N_cL/2}{E_cA_c} + \frac{N_{steel}L/2}{E_{steel}A_{steel}} = \alpha_{steel}\Delta T\frac{L}{2} - \alpha_c\Delta T\frac{L}{2} = (\alpha_{steel} - \alpha_c)\Delta T\frac{L}{2}$$

From the equilibrium of the composite cylinder:

$$N_c = N_{steel}$$

Thus:

$$N_c = N_{steel} = \frac{E_{steel}A_{steel}E_cA_c}{E_{steel}A_{steel} + E_cA_c}(\alpha_{Steel} - \alpha_c)\Delta T$$

The stresses in the concrete and steel are therefore:

$$\sigma_c = \frac{N_c}{A_c} = \frac{E_{steel}A_{steel}E_c}{E_{steel}A_{steel} + E_cA_c}(\alpha_{Steel} - \alpha_c)\Delta T$$

$$\sigma_{steel} = \frac{N_{steel}}{A_{steel}} = \frac{E_{steel}E_cA_c}{E_{steel}A_{steel} + E_cA_c}(\alpha_{Steel} - \alpha_c)\Delta T$$

2.12   Conceptual questions

1. Explain the terms 'stress' and 'strain' as applied to a bar in tension. What is the relationship between the two quantities?
2. What is meant by 'elastic material'? Define Modulus of elasticity.
3. Why does the axial stress vary along an axially loaded bar of variable section?
4. Consider the stresses in an axially loaded member. Which of the following statements are correct?

    (a) On the section where the maximum tensile stress occurs, shear stress vanishes.
    (b) On the section where the maximum tensile stress occurs, shear stress exists.
    (c) On the section where the maximum shear stress occurs, normal stress vanishes.
    (d) On the section where the maximum shear stress occurs, normal stress may exist.

5. A section subjected to maximum axial force is the section that will fail first. Is this correct, and why?
6. Two bars that are geometrically identical are subjected to the same axial loads. The bars are made of steel and timber, respectively. Which of the following statements is correct? The bars have

    (a) the same stresses and strains;
    (b) different stresses and strains;

    (c) the same stresses but different strains;

    (d) the same strains but different stresses.

7.   Which following statements are correct about elastic modulus E?

    (a) E represents the capacity of a material's resistance to deformation.

    (b) E is a material constant and is independent of stress level in a material.

    (c) E depends on cross-sectional area when a member is in tension.

    (d) E depends on the magnitude and direction of the applied load.

8.   What is your understanding of Saint-Venant's principle? And what is the importance of the principle in relation to static stress analysis?

9.   Two identical bars are subjected to axial tension as shown in Figure Q2.9. If the normal stress on the cross-section at the middle of both bars are the same,

    (a) what is the relationship between the concentrated force P and the uniformly distributed end pressure p?

    (b) is the distribution of normal stress on the sections near the ends of the bars the same as the distribution on the midsection and why?

Figure Q2.9

10.   What is meant by 'stress concentration'? Which of the bars shown in Figure Q2.10 is more sensitive to stress concentration?

Figure Q2.10

11.   Three bars are made of the same material and the shapes of their cross-section are, respectively, solid square, solid circle and hollow circle. If the bars are subjected to the same axial force, which of the following statements is correct?

    (a) To achieve the same level of stiffness, the square section uses less material.

    (b) To achieve the same level of stiffness, the circular section uses less material.

    (c) To achieve the same level of stiffness, the hollow circular section uses less material.

    (d) To achieve the same level of stiffness, the three sections use the same amount of material.

12. The symmetric space truss consists of four members of equal length (Figure Q2.12). The relationship between the axial stiffness of the members is $E_1A_1 > E_2A_2 > E_3A_3 > E_4A_4$. If a vertical force F is applied at the joint of the four members, in which member does the maximum axial force occur and why?

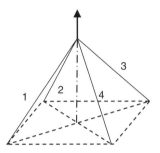

Figure Q2.12

13. The following members are subjected to a uniform change in temperature. In which members will the temperature change cause stresses in the axial direction?

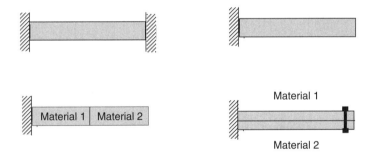

Figure Q2.13

2.13 Mini test

Problem 2.1: The four bars shown in the figure are made of different materials and have an identical cross-sectional area A. Can the normal stresses on the sections in middle span be calculated by P/A, and why?

Figure P2.1

Problem 2.2:   The bar with variable section is subjected to an axial force *P* as shown in Figure P2.2.

Figure P2.2

The total elongation, the strain and the total strain energy of the bar are calculated as follows:

Total elongation $\quad \Delta l = \Delta l_1 + \Delta l_2 = \dfrac{2Pl_1}{EA_1} + \dfrac{Pl_2}{EA_2}$

Strain $\quad\quad\quad\quad \varepsilon = \dfrac{\Delta l_1}{l_1} + \dfrac{\Delta l_2}{l_2}$

Total strain energy $\quad U = U_1 + U_2 = \dfrac{2P^2 l_1}{2EA_1} + \dfrac{P^2 l_2}{2EA_2}$

Are the calculations correct, and why?

Problem 2.3:   A force of 500 kN is applied at joint B to the pin-joined truss as shown in Figure P2.3. Determine the required cross-sectional area of the members if the allowable stresses are 100 MPa in tension and 70 MPa in compression.

Figure P2.3

Problem 2.4:   The members of the pin-joined truss shown in Figure P2.4 have the same tensional stiffness *EA*. Determine the axial forces developed in the three members due to the applied force *P*.

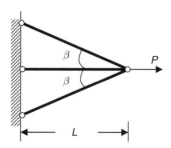

Figure P2.4

Problem 2.5:   A weight of 30 kN is supported by a short concrete column of square section 250 mm×250 mm. The column is strengthened by four steel bars in the corners of total cross-sectional area 6×10³ mm². If the modulus of elasticity for steel is 15 times that for concrete, find the stresses in the steel and the concrete.

250 mm

250 mm

Figure P2.5

If the stress in the concrete must not exceed 400 kN/m², what area of steel is required in order that the column may support a load of 60 kN?

3 Torsion

Torsion is another basic type of deformation of a structural member that is subjected to a twist action of applied forces, as shown by the cantilever shaft subjected to a torque at the free end (Figure 3.1). If the shaft is long and has a circular section, its torsion and deformation are characterized by the following:

Figure 3.1

- The torque or twist moment is applied within a plane perpendicular to the axis of the circular member.
- Under the action of the torque, shear stress develops on the cross-sections.
- Under the action of the torque, the deformation of the bar is dominated by angle of twist, i.e., the relative rotation between parallel planes perpendicular to the axis.
- A plane section perpendicular to the axis remains plane after the twist moment is applied, i.e., no warpage or distortion of parallel planes normal to the axis of a member occurs.
- In a circular member subjected to torsion, both shear stresses and shear strains vary linearly from the central axis.

3.1 Sign convention

A positive torque is a moment that acts on the cross-section in a right-hand-rule sense about the outer normal to the cross-section (Figure 3.2). Consequently, a positive angle of twist is a rotation of the cross-section in a right-hand-rule sense about the outer normal. As defined from the sign convention for the normal forces in Chapter 2, the sign convention defined here is again not related to a particular direction of coordinates.

Figure 3.2

3.2   Shear stress

Stresses that are developed on a cross-section due to torsion are parallel to the section and, therefore, are shear stresses. In a circular member subjected to a torque, shear strain, γ, varies linearly from zero at the central axis. By Hooke's law (Equation (1.5)), the shear stress, τ, is proportional to shear strain, that is:

$$\tau = G\gamma \qquad\qquad (3.1)$$

where G denotes shear modulus of material.

On a cross-section, shear stress also varies linearly from the central axis (Figure 3.3):

$$\tau = \frac{Tr}{J} \qquad\qquad (3.2)$$

where

T – torque acting on the section
r – radial distance from the centre
J – polar moment of inertia or polar second moment of area, representing a geometric quantity of the cross-section and having a unit of, for example, m⁴. The mathematical expression of J is:

$$J = \int_{area} r^2 dA(\text{rea}) \qquad\qquad (3.3)$$

(a)   (b)

Figure 3.3

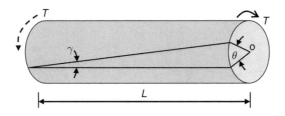

Figure 3.4

For the tubular section shown in Figure 3.3(b):

$$J = \frac{\pi}{32}(D_{out}^4 - D_{in}^4)$$ (3.4a)

where D_{out} and D_{in} are, respectively, the outside and inside diameters of the section. For the solid section shown in Figure 3.3(a), the inner diameter (D_{in}) equals zero, that is:

$$J = \frac{\pi}{32}D_{out}^4$$ (3.4b)

3.3   Angle of twist

Angle of twist is the angle difference between two parallel sections of a bar subjected to torsion. It is proportional to the applied torsion, T, and the distance between the two sections, L, while inversely proportional to the geometric quantity of cross-section, J, and the shear modulus, G, where GJ is called torsional rigidity. Thus for the bar shown in Figure 3.4:

$$\theta = \frac{TL}{GJ}$$ (3.5)

If the changes in twist moment, cross-sectional geometry and shear modulus along the central axis between sections are discrete, the total angle of twist is:

$$\theta = \sum_{i=1}^{N} \frac{T_i L_i}{G_i J_i}$$ (3.6)

where N is the total number of the discrete segments ($i = 1, 2, \ldots, N$), within each of which T_i, G_i and J_i are all constant.

If the changes are continuous for T, J and G within length L:

$$\theta = \int_0^L \frac{T(x)}{G(x)J(x)}dx$$ (3.7)

3.4   Torsion of rotating shafts

Members as rotation shafts for transmitting power are usually subjected to torque. The following formula is used for the conversion of kilowatts (kW), a common unit used in the industry, into torque applied on a shaft:

$$T = 159\frac{p}{f}$$

or

$$T = 9540\frac{p}{N}$$ (3.8)

Where

T = torque in N m
p = transmitted power in kW
f = frequency of rotating shaft in Hz
N = revolutions per minutes of rotating shaft (rpm).

3.5 Key points review

- If a shaft is long and the loads applied are twist moments/torques about the longitudinal axis, the deformation can be defined by the angle of twist.
- The internal stresses and forces on internal cross-sections perpendicular to the axis are shear stresses and torques, respectively.
- The resultant of the shear stresses on any cross-section along the shaft is the torque acting on the same section.
- The polar second moment of area and the shear modulus represent, respectively, the geometric and material contributions of the shaft to the torsional stiffness.
- The torsional stiffness of a section is GJ, which defines torsional resistance of a member.
- Shear stress is proportional to shear strain and varies linearly from the central axis.
- On a cross-section, maximum shear stress always occurs at a point along the outside boundary of a section.
- Equations (3.2)–(3.7) do not apply for noncircular members, for which cross-sections perpendicular to the axis warp when a torque is applied.

3.6 Recommended procedure of solution

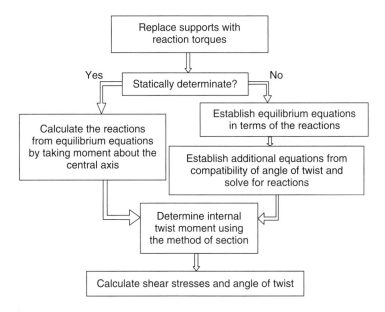

3.7 Examples

EXAMPLE 3.1

The solid shaft shown in Figure E3.1(a) is subjected to two concentrated torques, respectively, at A and B. The diameter of the shaft changes from 1 m to 0.5 m at B. What is the
total angle of twist at locations A and B and what is the maximum shear stress within the
shaft? $E = 200\,\text{GPa}$ and $\nu = 0.25$.

[Solution] *This is a statically determinate system subjected to only twist moments. The reaction
torque at the built-in end can be determined from the equilibrium of moment. The method of
section can be used to calculate the torques on the sections between A and B and B and C. The
stresses on these sections can then be calculated by Equation (3.2). Because the shaft is under a
discrete variation of torques and geometrical dimension, Equation (3.6) is used to calculate the
angle of twist.*

Figure E3.1(a)

Figure E3.1(b)

From the equilibrium of Figure E3.1(b):

$$T_c - T_A - T_B = 0$$

$$T_c = T_A + T_B = 5\,\text{MNm}$$

By taking sections between A and B, and B and C (see Figure E3.1(b)), the twist moment diagram
is drawn in Figure E3.1(c). Following the right-hand rule, the twist moments on the sections
along the axis are defined as positive.
Total angle of twist:

$$\text{Since } G_{AB} = G_{BC} = G = \frac{E}{2(1+\nu)} = \frac{200}{2(1+0.25)} = 80\,\text{GPa}$$

$$J_{AB} = \frac{\pi}{32}(0.5)^4 = \frac{\pi}{512}\,\text{m}^4$$

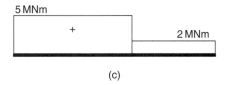

(c)

Figure E3.1(c)

$$J_{BC} = \frac{\pi}{32}(1)^4 = \frac{\pi}{32}m^4$$

From Equation (3.6), the angle of twist at A is:

$$\theta_A = \sum_{i=1,2} \frac{T_i L_i}{G_i J_i} = \frac{T_{AB}L_{AB}}{G_{AB}J_{AB}} + \frac{T_{BC}L_{BC}}{G_{BC}J_{BC}}$$

$$= \frac{2 \times 4}{80 \times 10^3 \pi/512} + \frac{5 \times 5}{80 \times 10^3 \pi/32} = 19.48 \times 10^{-3}(\text{rad})$$

The angle of twist at B is:

$$\theta_B = \frac{T_{BC}L_{BC}}{G_{BC}J_{BC}} = \frac{5 \times 5}{80 \times 10^3 \pi/32} = 3.18 \times 10^{-3}(\text{rad})$$

Maximum shear stress
The maximum shear stresses within AB and BC are needed for the overall maximum.
From Equation (3.2):

$$\tau_{AB}^{max} = \frac{T_{AB}r_{max}}{J_{AB}} = \frac{2 \times (0.5/2)}{\pi/512} = 81.49\,\text{MPa}$$

$$\tau_{BC}^{max} = \frac{T_{BC}r_{max}}{J_{BC}} = \frac{5 \times (1/2)}{\pi/32} = 25.46\,\text{MPa}$$

Thus, the maximum shear stress occurs between A and B.

EXAMPLE 3.2
A motor drives a circular shaft through a set of gears at 630 rpm; 20 kW are delivered to a machine on the right and 60 kW on the left (Figure E3.2). If the allowable shear stress of the shaft is 37 MPa, determine the minimum diameter of the shaft.

Figure E3.2

[Solution] *The torsion problem is statically determinate. The torques acting on the shaft have to be calculated first by Equation (3.8). The shaft is then analyzed by Equation (3.2) to determine the maximum shear stress that depends on the diameter. A comparison of the maximum shear stress with the allowable stress of the material leads to the solution of the minimum diameter.*

Since the total power transmitted to the shaft through gear B is 80 kW (60 + 20):

$$T_B = 9540\frac{P}{N} = 9540 \times \frac{80}{630} = 1211.43\,\text{N m}$$

The resistance torques at A and C are, respectively:

$$T_A = 9540\frac{P}{N} = 9540 \times \frac{20}{630} = 302.86\,\text{N m}$$

$$T_C = 9540\frac{P}{N} = 9540 \times \frac{60}{630} = 908.57\,\text{N m}$$

The twist moment diagram of the shaft is shown below:

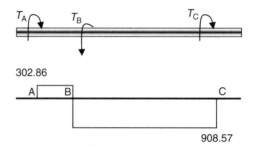

302.86

908.57

Because the shaft has a constant diameter, it is obvious that the maximum twist moment, then the maximum shear stress, occurs between B and C. By Equation (3.2):

$$\tau_{max} = \frac{T_{max}D/2}{J}$$

$$= \frac{908.57(\text{N m}) \times D/2}{\pi D^4/32} \leq \tau_{allowable} = 37 \times 10^6\,\text{N/m}^2$$

$$D \geq 5 \times 10^{-2}\,\text{m} = 50\,\text{mm}$$

Thus, the diameter of the shaft must not be smaller than 50 mm.

EXAMPLE 3.3
A uniform steel pile of circular section, which has been driven to a depth of L in clay, carries an applied torque T at the top. This load is resisted entirely by the moment of friction m_f along the pile, which is linearly distributed along the depth as shown in Figure E3.3(a). If $L = 60\,\text{m}$, the diameter of the pile $D = 100\,\text{mm}$, $T = 4\,\text{kN m}$, shear modulus $G = 80\,\text{GPa}$ and the allowable shear stress of material $\tau_{allowable} = 40\,\text{MPa}$. Check the safety of the pile and determine the total angle of twist.

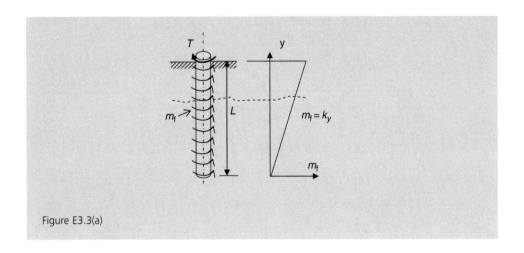

Figure E3.3(a)

[Solution] *This is a statically determinate system. From the equilibrium requirement, the applied torque must be equal to the resultant torque due to the moment of friction, from which the constant k can be determined. The method of section can be used to find the torque on an arbitrary section and then draw the diagram of torque along the axis of the pile. The units used in the following calculation are m and N.*

From the equilibrium of the pile, the resultant of m_f must equal the applied torque T. Thus, from Figure E3.3(a):

$$T = \text{Area of the triangle} = \frac{60 \times 60 k}{2} = 4000 \, \text{N m}$$

$$k = \frac{4000 \times 2}{60 \times 60} = 2.22 \, \text{N m/m}$$

Take a cut at y (Figure E3.3(b)) and consider equilibrium of the low segment.

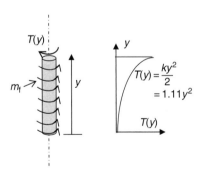

Figure E3.3(b)

$$T(y) = \frac{ky \times y}{2} = 1.11y^2$$

The maximum torque occurs on the section at $y = 60$ m, that is, the section in level with the ground surface:

$$\tau_{max} = \frac{T(L) \times D/2}{J} = \frac{4 \times 10^3 \times 50 \times 10^{-3}}{\frac{\pi}{32} \times 100^4 \times 10^{-12}}$$

$$= 20.37 \, \text{MPa} < 40 \, \text{MPa}$$

The design of the pile is satisfactory. Since the pile is subjected to the continuously distributed twist moment shown in Figure E3.3(b), from Equation (3.7):

$$\theta_{total} = \int_0^L \frac{T(y)}{G(y)J(y)} dy = \int_0^L \frac{ky^2/2}{GJ} dy = \frac{k}{2GJ} \int_0^L y^2 dy$$

$$= \frac{kL^3}{6GJ}$$

$$= \frac{1.11 \times 60^3}{6 \times 80 \times 10^9 \times \frac{\pi}{32} \times 100^4 \times 10^{-12}}$$

$$= 0.051 \, (\text{rad}) = 2.92^0$$

EXAMPLE 3.4

A bar of length 4 m shown in Figure E3.4(a) has a circular hollow cross-section with an outside diameter D_{out} of 50 mm and is rigidly built in at each end. It carries torques of 0.9 and 1.5 kN m at the mid-span and three-quarter span sections, respectively, taken from the left-hand end. If both torques are applied in the same direction and the maximum shear stress in the bar is limited to $100 \, \text{N/mm}^2$, calculate the maximum allowable internal diameter D_{in} of the bar.

[Solution] *This is a statically indeterminate structure and hence the reaction torques on the cross-sections at A and D cannot be uniquely determined by considering the equilibrium of the system. Thus a compatibility requirement must be considered, for example, by replacing the support at D with a twist moment T_D (Figure 3.4(a)). Under the action of the three twist moments, the angle of twist of section D must be zero (fixed end condition). After the end moment at D is found, the system is equivalent to a statically determinate system. The procedure shown in previous examples can then be followed to determine the final solution.*

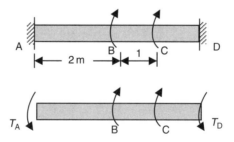

Figure E3.4(a)

Assume that the reaction torques acting on the cross-sections at A and D are, respectively, T_A and T_D. From the equilibrium of the bar:

$$T_A + T_D - T_B - T_C = 0$$

$$T_A + T_D = T_B + T_C = 2.4 \, \text{kN m}$$

Now consider the angle of twist at D under the combined action of T_B, T_C and T_D, i.e., as shown in Figure E3.4(b):

Figure E3.4(b)

as a superimposition of the angle of twist caused by the action of these torques individually, as shown in Figures E3.4(c), (d) and (e), respectively.

From Figure E3.4(c) and Equation (3.5):

Figure E3.4(c)

$$(\theta_D)_{(c)} = \theta_B = \frac{T_B L_{AB}}{GJ}$$

From Figure E3.4(d) and Equation (3.5):

Figure E3.4(d)

$$(\theta_D)_{(d)} = \theta_C = \frac{T_C L_{AC}}{GJ}$$

And from Figure E3.4(e) and Equation (3.5):

Figure E3.4(e)

$$(\theta_D)_{(e)} = \frac{T_D L_{AD}}{GJ}$$

The compatibility condition requires:

$$(\theta_D)_{(c)} + (\theta_D)_{(d)} - (\theta_D)_{(e)} = 0$$

Thus:

$$\frac{T_B L_{AB}}{GJ} + \frac{T_C L_{AC}}{GJ} - \frac{T_D L_{AD}}{GJ} = 0$$

$$0.9 \times 2 + 1.5 \times 3 - T_D \times 4 = 0$$

$$T_D = 1.57 \, \text{kN m}$$

From the equilibrium:

$$T_A = 2.4 - T_D = 2.4 - 1.575 = 0.825 \, \text{kN m}$$

From the twist moment diagram shown in Figure E3.4(f), the maximum twist moment occurs between C and D, where the design requirement must be met.

Figure E3.4(f)

Now from Equation (3.2):

$$\tau = \frac{Tr}{J} = \frac{1.575 \times 10^6 \times D_o/2}{\pi(D_{out}^4 - D_{in}^4)/32}$$

Thus, for the minimum internal diameter:

$$\frac{1.575 \times 10^6 \times D_{out}/2}{\pi(D_{out}^4 - D_{in}^4)/32} \leq 100$$

$$D_{in} = 38.7 \, \text{mm}$$

The internal diameter of the shaft must not be larger than 38.7 mm.

EXAMPLE 3.5
A tube of aluminium with a solid core of steel on the inside is shown in Figure E3.5. The member is subjected to a torque T. The shear moduli for aluminium and steel are, respectively, G_a and G_{steel}. What are the maximum shear stresses in the aluminium and the steel?

[Solution] *This is also a statically indeterminate problem. Due to the composite section, the shear stress distribution over the steel and aluminium areas has to be considered separately. Apart from the equilibrium condition, the compatibility of deformation condition is that the angle of twist has the same value for both the steel and the aluminium.*

Figure E3.5

Assume that the twist moments carried by the steel and the aluminium are, respectively, T_{steel} and T_a. Thus, from equilibrium:

$$T_a + T_{steel} = T$$

Since the steel and the aluminium have the same angle of twist, from Equation (3.5):

$$\frac{T_{steel}L}{G_{steel}J_{steel}} = \frac{T_a L}{G_a J_a}$$

From the above two equations:

$$T_a = \frac{G_a J_a}{G_a J_a + G_{steel}J_{steel}} T$$

$$T_{steel} = \frac{G_{steel}J_{steel}}{G_a J_a + G_{steel}J_{steel}} T$$

Then, the maximum shear stresses in the steel and the aluminium are, respectively:

$$\tau_a^{max} = \frac{T_a \times D/2}{J_a} = \frac{TG_a D}{2(G_a J_a + G_{steel}J_{steel})}$$

$$\tau_{steel}^{max} = \frac{T_{steel} \times d/2}{J_{steel}} = \frac{TG_{steel}d}{2(G_a J_a + G_{steel}J_{steel})}$$

EXAMPLE 3.6
A hollow steel shaft with an internal diameter of $d = 8$ in. and an outside diameter of $D = 12$ in. is to be replaced by a solid alloy shaft. If the maximum shear stress has the same value in both shafts, calculate the diameter of the latter and the ratio of the torsional rigidities $GJ \cdot G_{steel} = 2.4G_{alloy}$.

[Solution] *Because the maximum shear stress must be the same in both shafts under the same torque, from Equation (3.2), $[r_{max}/J]_{steel}$ must be equal to $[r_{max}/J]_{alloy}$.*

Thus:

$$\frac{D_{steel}/2}{J_{steel}} = \frac{D_{alloy}/2}{J_{alloy}}$$ (i)

For the shafts:

$$J_{steel} = \frac{\pi}{32}(12^4 - 8^4) = 520\pi \text{ in.}^4$$

$$J_{alloy} = \frac{\pi}{32}D_{alloy}^4$$

From (i):

$$\frac{6}{520\pi} = \frac{D_{alloy}/2}{\frac{\pi}{32}D_{alloy}^4}$$

Hence, $D_{alloy}^3 = 1386 \text{ in.}^3$ and $D_{alloy} = 11.15 \text{ in.}$
Ratio of torsional rigidity:

$$\frac{G_{steel}J_{steel}}{G_{alloy}J_{alloy}} = \frac{G_{steel}}{G_{alloy}} \times \frac{D_{steel}}{D_{alloy}} \qquad \text{[from (i)]}$$

$$= 2.4 \times \frac{12}{11.15} = 2.58$$

That is, the torsional rigidity of the steel is 2.58 times that of the alloy shaft. This ratio means that, though replacing the steel shaft by the alloy one meets the strength requirement, the angle of twist of the alloy shaft will be 2.58 times larger than that of the steel.

EXAMPLE 3.7
When a bar of rectangular section is under torsion (Figure E3.7), determine the shear stresses at the corners of the cross-section.

[Solution] *This question tests your conceptual understanding of shear stresses. They always appear in pair on planes that are perpendicular to each other. They are equal in magnitude, but in an opposite sense.*

Figure E3.7

Take an infinitesimal element around a corner point and assume that on the plane of the cross-section there are two shear stresses perpendicular to either the horizontal or the vertical boundary. The two assumed shear stresses must be equal to the shear stresses acting on the top and the left-hand-side surfaces. Since the bar has a stress-free surface, the surface shear stresses are zero (Section 1.4.2). Consequently, the two assumed shear stresses at the corner of the cross-section are also zero.

3.8 Conceptual questions

1. Explain the terms 'shear stress' and 'shear strain' as applied to the cross-section of a long shaft subjected to torsion.
2. Explain the terms 'torsional rigidity' and 'polar moment of inertia'.
3. Explain why Equations (3.2)–(3.7) can only apply for circular members.
4. Which of the following structures is under torsion?

(a)　　　　　　　(b)　　　　　　　(c)

Figure Q3.4

5. When a circular cross-section is under pure torsion, why does the direction of the shear stress along the circular boundary always coincide with the tangent of the boundary?
6. When a twist moment is applied at a location along the axis of a bar, is it right to say that the twist moment on the cross-section at the same location is equal to the external applied moment and why?
7. The twist moment diagrams for a shaft subjected to torsion are shown in Figure Q3.7. Determine the magnitude, direction and location of the externally applied moments for each case.

Figure Q3.7

8. The tube sections below have the same outside radius and are subjected to the same twist moment.

Figure Q3.8

Which of the following statements are correct and why?

(a) The maximum shear stresses are the same on both sections.
(b) The maximum shear stress of (a) is greater than that of (b).
(c) The maximum shear stress of (b) is greater than that of (a).
(d) The minimum shear stress of (b) is greater than that of (a).

9. Under torsion, the small square in Figure Q3.9 will change to

Figure Q3.9

(a) a larger square
(b) a rectangle
(c) a diamond
(d) a parallelogram.

10. Brittle materials normally fail at maximum tension. If the shaft shown in Figure Q3.10 is made of iron and subjected to pure torsion, which of the following statements is correct?

Figure Q3.10

(a) The shaft is likely to fail along section 1-1.
(b) The shaft is likely to fail along section 2-2.
(c) The shaft is likely to fail along section 3-3.
(d) The shaft is likely to fail along section 4-4.

11. A solid circular section of a diameter D can carry a maximum torque T. If the circular area of the cross-section is doubled, what is the maximum torque that the new section can carry?

(a) $\sqrt{2}T$
(b) $2\sqrt{2}T$
(c) $2T$
(d) $4T$

12. A solid circular section has the same area as a hollow circular section. Which section has higher strength and stiffness, and why?

13. A solid circular shaft of steel and a solid circular shaft of aluminium are subjected to the same torque. If both shafts have the same angle of twist per unit length (θ/L), which one of the following relationships will the maximum shear stresses in the shafts satisfy?

(a) $\tau_{steel} < \tau_{aluminium}$
(b) $\tau_{steel} > \tau_{aluminium}$
(c) $\tau_{steel} = \tau_{aluminium}$
(d)   All of the above are possible.

3.9   Mini test

Problem 3.1: The twist moment diagrams are shown below for a shaft subjected to torsion. Determine the magnitude, direction and location of the externally applied moments for each case.

Figure P.3.1

Problem 3.2: Two geometrically identical shafts are loaded with the same twist moments. If the two shafts are made of different materials, which of the following statements is correct.

(a)   Both the maximum shear stresses and the angles of twist are the same.
(b)   Both the maximum shear stresses and the angles of twist are different.
(c)   The maximum shear stresses are the same, but the angles of twist are different.
(d)   The maximum shear stresses are different, but the angles of twist are the same.

Problem 3.3: A hollow circular section is subjected to torsion. Which of the following stress distributions is correct?

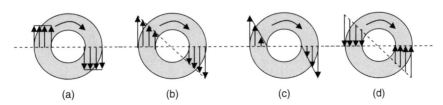

(a)   (b)   (c)   (d)

Figure P3.3

Problem 3.4: Calculate the minimum diameters of two shafts transmitting 200 kW each without exceeding the allowable shear stress of 70 MPa. One of the shafts rotates at 20 rpm and the other at 20 000 rpm. What conclusion can you make from the relationship between the stresses and the speed of the shafts?

Problem 3.5: Consider the stepped shaft shown in Figure P3.5. The shaft is fixed at both ends. Assuming that a, d, G and T are all given constants, determine the maximum shear stress and the angle of twist at B.

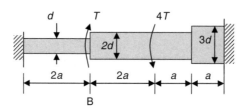

Figure P3.5

4 Shear and bending moment

Figure 4.1(a) shows a car crossing a bridge. The purpose of the bridge deck is to transfer the weight of the car to the two supports. Figure 4.1(b) is an equivalent illustration of Figure 4.1(a), showing the transversely applied load (the car), the structure (the deck) and the supports at the ends. In fact, in many engineering applications, structural members resist forces (loads) applied laterally or transversely to their axis. This type of member is termed a *beam*. The deformation of a beam is characterized as follows:

- The external load is applied transversely and causes the beam to flex as shown in Figure 4.1(c).
- Bending moment and shear force develop on cross-sections of the beam as shown in Figure 4.1(d).
- Under the action of the bending moment and shear force, the deformation of the beam is dominated by transverse deflection and a rotation of cross-section.
- Axial deformation of the beam is neglected.

(a)

(b)

(c)

(d)

Figure 4.1

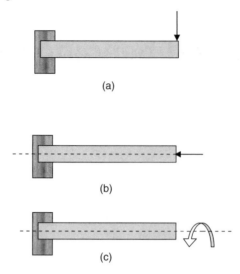

(a)

(b)

(c)

Figure 4.2

4.1 Definition of beam

A beam is defined as a structural member designed primarily to support forces acting perpendicular to the axis of the member (Figure 4.2). The major difference between beams (Figure 4.2(a)) and axially loaded bars (Figure 4.2(b) or shafts in torsion (Figure 4.2(c)) is in the direction of the applied loads.

4.2 Shear force and bending moment

A shear force is an internal force that is parallel to the section it is acting on and that, for example, in Figure 4.1(d), resists the vertical effect of the applied loads on the beam. The shear force is numerically equal to the algebraic sum of all the vertical forces acting on the free body taken from either sides of the section. Shear forces are measured in N, kN, etc.

A bending moment is an internal force that resists the effect of moments caused by external loads, including support reactions. Bending moments are measured in N m, kN m, etc.

4.3 Beam supports

A beam may be supported differently. Table 4.1 shows the most common types of supports that are frequently used to model practical structural supports in design.

4.4 Sign convention

The following sign convention is adopted for bending analysis of beams in most textbooks. The convention is again not in relation to a particular coordinate direction.

4.4.1 Definition of positive shear

A downward shear force acting on the cross-section of the left-hand-side free body of a cut, or an upward shear force acting on the cross-section of the right-hand-side free body of the cut, is defined as a positive shear force (Figure 4.3(a)). Positive shear forces are shown in Figure 4.3(b) for a segment isolated from a beam by two sections (cuts).

Table 4.1 Beam supports and reactions

Type of support	Simple illustration	Reaction forces and displacements at the support	
Roller	or	(1) Shear force	(1) Axial displacement (usually ignored*) (2) Rotation
Pinned		(1) Shear force (2) Axial force (usually ignored*)	(1) Rotation
Fixed		(1) Shear force (2) Axial force (usually ignored*) (3) Bending moment	None
Free		None	(1) Axial displacement (usually ignored*) (2) Vertical displacement (3) Rotation

* Axial force/displacement is significantly smaller than other types of forces/displacements in bending.

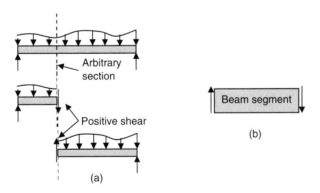

Figure 4.3

The shear forces shown in Figure 4.3(b) tend to *push up* the left-hand side of the beam segment. The positive shear forces can also be described as *left up* shear forces.

4.4.2 Definition of positive bending moment

A positive bending moment (Figure 4.4(a)) produces compression in the upper part and tension in the lower part of a beam's cross-section. The deformed beam takes a shape that can *retain water*.

From Figure 4.4(b), a positive bending moment can also be described as *sagging moment* since the moment induces a sagging deflection.

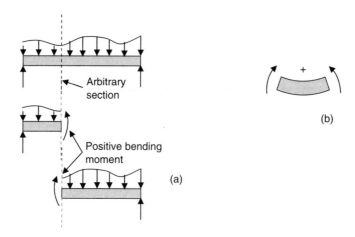

Figure 4.4

4.5   Relationships between bending moment, shear force and applied load

On the arbitrary cross-section shown in Figure 4.5, the interrelation of the bending moment $M(x)$, shear force $V(x)$ and the intensity of the distributed load $q(x)$ always obey the following relationships.

(i)   The change rate of shear force along a beam is equal to the distributed load:

$$\frac{dV(x)}{dx} = -q(x) \tag{4.1}$$

(ii)   The change rate of bending moment along a beam is equal to the shear force:

$$\frac{dM(x)}{dx} = V(x) \tag{4.2}$$

(iii)   The combination of Equations (4.1) and (4.2) yields:

$$\frac{d^2M(x)}{dx^2} = -q(x) \tag{4.3}$$

The above relationships are applicable at all cross-sections of a beam except where there is a concentrated force or moment.

Figure 4.5

4.6   Shear force and bending moment diagrams

The shear force and bending moment diagrams indicate, respectively, the internal shear force and bending moment distribution on the cross-section along the length of a beam.

On the basis of the interrelations shown in Section 4.5, Table 4.2 shows some common features exhibited in the shear force and bending moment diagrams of beams.

Table 4.2 Common features of shear and bending moment diagrams

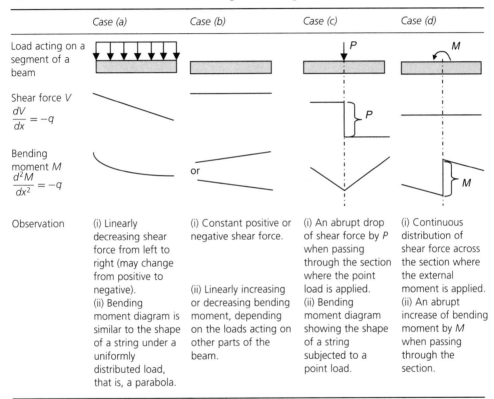

	Case (a)	Case (b)	Case (c)	Case (d)
Load acting on a segment of a beam			P	M
Shear force V $\dfrac{dV}{dx} = -q$			P	
Bending moment M $\dfrac{d^2M}{dx^2} = -q$		or		M
Observation	(i) Linearly decreasing shear force from left to right (may change from positive to negative). (ii) Bending moment diagram is similar to the shape of a string under a uniformly distributed load, that is, a parabola.	(i) Constant positive or negative shear force. (ii) Linearly increasing or decreasing bending moment, depending on the loads acting on other parts of the beam.	(i) An abrupt drop of shear force by P when passing through the section where the point load is applied. (ii) Bending moment diagram showing the shape of a string subjected to a point load.	(i) Continuous distribution of shear force across the section where the external moment is applied. (ii) An abrupt increase of bending moment by M when passing through the section.

4.7   Key points review

- A beam is long in the axial direction compared to its other two dimensions.
- A beam is to support external load applied perpendicular to the axis.
- The two major internal forces are shear force and bending moment.
- The change in shear force is equal to applied distributed load.
- The change in bending moment is equal to shear force.
- A shear force diagram shows how shear force is distributed along the axis of a beam.
- A bending moment diagram shows how bending moment is distributed along the axis of a beam.
- Continuous distribution of applied loads result in continuous variations of both shear force and bending moment along a beam.
- On unloaded segments of a beam the shear force is constant.
- A uniformly distributed load causes a linear variation in shear force, and a parabolic variation in bending moment.

- A concentrated force/moment results in sudden drop or jump of shear force/bending moment at the location where the force/moment is applied.
- A point force causes a *kink* in a bending moment diagram.
- A concentrated moment has no effect on a shear force diagram.
- There is no shear force at a free end of a beam.
- There is no bending moment at a free or simply supported end of a beam.
- An internal hinge can transmit a shear force.
- An internal hinge cannot transmit a bending moment.
- Bending moment reaches either maximum or minimum at a point of zero shear force.

4.8   Recommended procedure of solution

- Replace all supports of a beam by their associated reactions (see Table 4.1).
- Apply the static equilibrium equations to determine the reactions.
- Identify *critical sections* that characterize changes of pattern of the internal force diagrams. The critical sections include the locations where (a) a concentrated force or moment is applied; (b) a beam is supported; and (c) a distributed load starts or ends.
- Apply the method of section by taking cuts between each of the critical sections.
- Take either left or right part of the cut as a free body.
- Add the unknown shear force and bending moment on the cut of the free body.
- Consider static equilibrium of the free body and calculate the shear force and bending moment on the cross-section (cut).
- Use Table 4.2 to draw the diagrams between the critical sections. If a moment distribution is parabolic between two critical sections and exact distribution is required, at least an additional cut must be taken between the two sections.

4.9   Examples

EXAMPLE 4.1

Draw the shear force and bending moment diagrams of the simply supported beam shown in Figure E4.1.

Figure E4.1

[Solution] *This is a determinate problem that can be solved by following the general procedure described in Section 4.8.*

Step 1: Replace supports by reactions.

Since there is no external load applied horizontally, the horizontal support reaction at A vanishes. Thus the pin at A is replaced by a single reaction R_A in the vertical direction.

Step 2: Solution of the reactions.

Taking moments about A ($\stackrel{\frown}{A}$):

$$\sum M_A = 10\,\text{m} \times R_B + 2\,\text{kN} \times 4\,\text{m} + 2\,\text{kN m} - 4\,\text{kN} \times 8\,\text{m} - 5\,\text{kN/m} \times 2\,\text{m} \times 1\,\text{m} = 0$$

$$R_F = 3.2\,\text{kN}(\uparrow)$$

Resolving vertically ($\uparrow$):

$$R_A + R_F + 2\,\text{kN} - 4\,\text{KN} - 5\,\text{kN/m} \times 2\,\text{m} = 0$$

$$R_A = 8.8\,\text{kN}(\uparrow)$$

Step 3: Identify critical sections.

Sections at A, B, C, D, E and F are all critical sections, where either application of concentrated forces or change of load pattern occurs.

Step 4: Calculation of shear forces and bending moments on the critical sections.

(a) *Section at A*

At section A (the end section), R_A acts as a positive shear force (left up), while there is no bending moment. Thus:

$$V_A = R_A = 8.8\,\text{kN}$$

$$M_A = 0\,\text{kN}$$

(b) *Section at B*

Resolving vertically:

$$R_A - 5\,\text{kN/m} \times 2\,\text{m} - V_B = 0$$

$$V_B = -1.2\,\text{kN}(\uparrow)$$

The negative sign denotes that the actual direction of V_B is opposite to the assumed direction.

Taking moment about B ($\overset{\frown}{B}$):

$$M_B + 5\,\text{kN/m} \times 2\,\text{m} \times 1\,\text{m} - R_A \times 2\,\text{m} = 0$$

$$M_B = 7.6\,\text{kN m}\left(\overset{\curvearrowleft}{}\right)$$

In order to draw the parabolic distribution of the bending moment between A and B, an additional section between A and B must be considered. Let us take the middle span of AB.

Taking moment about the cut:

$$M_{AB} + 5\,\text{kN/m} \times 1\,\text{m} \times 0.5\,\text{m} - R_A \times 1\,\text{m} = 0$$

$$M_{AB} = 6.3\,\text{kN m}\left(\overset{\curvearrowleft}{}\right)$$

(c) *Section at C*

We can consider the section on the immediate left or right of the concentrated load applied at C, that is, with or without including the load in the free body diagram. We take the section on the left of the load at C in the following calculations.

Resolving vertically:

$$R_A - 5\,\text{kN/m} \times 2\,\text{m} - V_C = 0$$

$$V_C = -1.2\,\text{kN}(\uparrow)$$

Taking moment about C ($\stackrel{\frown}{C}$):

$$M_C + 5\,kN/m \times 2\,m \times 3\,m - R_A \times 4\,m = 0$$

$$M_C = 5.2\,kN\,m\left(\stackrel{\frown}{}\right)$$

(d)   *Section at D*
Once again, we can consider the section on either the immediate left or the immediate right of the concentrated moment applied at D, that is, with or without including the moment in the free body diagram. We take the section on the left of the moment at D in the following calculations.

Resolving vertically:

$$R_A + 2\,kN - 5\,kN/m \times 2\,m - V_D = 0$$

$$V_D = 0.8\,kN(\downarrow)$$

Taking moment about D ($\stackrel{\frown}{D}$):

$$M_D + 5\,kN/m \times 2\,m \times 5\,m - 2\,kN \times 2\,m - R_A \times 6\,m = 0$$

$$M_D = 6.8\,kN\,m\left(\stackrel{\frown}{}\right)$$

(e)   *Section at E*
We take the section on the left-hand side of the concentrated load applied at E.

Resolving vertically:

$$R_A + 2\,kN - 5\,kN/m \times 2\,m - V_E = 0$$

$$V_E = 0.8\,kN(\downarrow)$$

Taking moment about E ($\widehat{E}$):

$$M_E + 2\,\text{kN m} + 5\,\text{kN/m} \times 2\,\text{m} \times 7\,\text{m} - 2\,\text{kN} \times 4\,\text{m} - R_A \times 8\,\text{m} = 0$$

$$M_E = 6.4\,\text{kN m}\left(\curvearrowleft\right)$$

(f) *Section at F*

The only reaction at F is the vertically upward force R_F that is the shear force on the section. Thus:

$$V_F = R_F = -3.2\,\text{kN}$$
$$M_F = 0$$

The shear force is negative by following the sign convention established in Section 4.4.1.

Step 5: Shear force and bending moment diagrams.

From Table 4.2, the shear force diagram between A and B is a straight line valued 8.8 kN at A and −1.2 kN at B, while the bending moment diagram between A and B is a parabola (Case (a) in Table 4.2). We have known the values of the bending moment at three locations from A to B, that is, at A, B and the middle point of AB. The accurate shape of the parabola can then be drawn.

Between B and C (Case (b) in Table 4.2), the shear force is a constant that equals the shear force at B (there is no concentrated force applied here), that is, $V_B = -1.2\,\text{kN}$. The bending moment diagram is a straight line valued 7.6 kN m at B and 5.2 kN m at C (there is no concentrated moment applied at B). Since there are no concentrated force and moment applied at B, both the shear force and the bending moment show no abrupt changes across B.

Between C and D (Case (b) in Table 4.2), the shear force is a constant valued 0.8 kN (from $V_D = 0.8\,\text{kN}$), while the bending moment varies linearly with a value of 6.8 kN m at D. Obviously due to the concentrated force 2 kN applied at C, the shear force shows an abrupt jump in the direction of the applied force when crossing C from the left-hand side of the force (Case (c) in Table 4.2). Since there is no concentrated moment applied at C, the bending moment diagram is continuous across section C.

Between D and E (Case (b) in Table 4.2), the shear force and bending moment are, respectively, constant and linear. Since there is no concentrated load applied at D, the constant shear between D and E is equal to the shear force between C and D, that is, 0.8 kN. The bending moment at D has an abrupt jump of 2 kN m (Case (d) in Table 4.2) due to the concentrated moment applied at D. The moment then varies linearly and is 6.4 kN m at E ($M_E = 6.4\,\text{kN m}$).

Between E and F (Case (b) in Table 4.2), the shear force is constant with an abrupt drop from 0.8 kN to −3.2 kN at E due to the applied downward load of 4 kN (Case (c) in Table 4.2).

The bending moment diagram is linear between E and F. Since there is no concentrated moment applied at E, the bending moment continues across the section. The bending moment vanishes as $M_F = 0$ at the section supported by the roller pin.

On the basis of the calculations in Step 4 and the analysis in Step 5, the shear force and bending moment diagrams of the beam are drawn below.

The following are observations from the above diagrams:

(i)   When shear force is a constant over a span of a beam, the bending moment over the same span is a sloped straight line (see Equation (4.2)).
(ii) When shear force is a sloped straight line over a span of a beam, the bending moment over the same span is a parabola (see Equation (4.2)).
(iii) At a point where shear force diagram passes through zero, the bending moment is either maximum or minimum.

EXAMPLE 4.2
Draw the shear force and bending moment diagrams of the beam shown in Figure E4.2. The beam has a pin joint at B.

Figure E4.2

[Solution] *This is a statically determinate beam. The pin at B joints AB and BD. Thus the bending moment is zero at B. From Table 4.2 and Example 4.9, we know that the shear force is constant and the bending moment varies linearly from A to C since there is no external load applied within this range. The shear force and the bending moment are, respectively, linear and parabolic between C and D. An abrupt change of shear force occurs at C due to the concentrated support reaction.*

Step 1: Replace supports with reactions.

The horizontal reaction at A vanishes in this case since there is no external load applied horizontally.

Step 2: Calculate support reactions.

Consider BD and take moment about B ($\widehat{B}$), noting that at B the bending moment is zero due to the hinge.

$$-qa^2 + R_C a - qa \times (a + \frac{a}{2}) = 0$$

$$R_C = \frac{5qa^2}{2}(\uparrow)$$

Considering AD and resolving vertically:

$$R_A + R_C - qa = 0$$

$$R_A = qa - R_C = -\frac{3qa}{2}(\downarrow)$$

Consider AB and take moment about B ($\widehat{B}$):

$$-R_A a - M_A = 0$$

$$M_A = -R_A a = -(-\frac{3qa}{2})a = \frac{3qa^2}{2}\left(\uparrow\right)$$

Step 3: Identify critical sections.
Sections at A, C and D are the critical sections.

Step 4: Calculate shear forces and bending moments on the critical sections.

(a)   Section at A
      From the calculation in Step 2:

$$V_A = R_A = -\frac{3qa}{2}$$

$$M_A = \frac{3qa^2}{2}$$

(b)   Section at C
      Take the section on the left of the roller pin at C.

Resolving vertically:

$$R_A - V_C = 0$$

$$V_C = R_A = -\frac{3qa}{2}(\uparrow)$$

Taking moment about C ($\widehat{C}$):

$$M_C - M_A - R_A \times 2a = 0$$

$$M_C = M_A + 2aR_A = \frac{3qa^2}{2} + 2a \times (-\frac{3qa}{2}) = -\frac{3qa^2}{2}\left(\curvearrowleft\right)$$

(c) *Section at D*

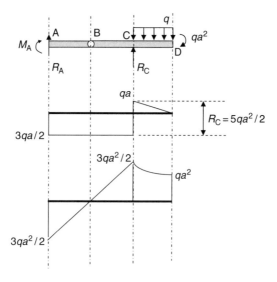

The section at D is a free end subjected to a concentrated moment. Thus:

$$V_D = 0$$
$$M_D = -qa^2$$

Following the sign convention, the moment at D is defined as negative.

Step 5: Draw shear force and bending moment diagrams.
On the basis of the calculations in Step 4 and the discussion before Step 1, the shear force and bending moment diagrams are drawn below.

The following are observations from the above shear force and bending moment diagrams:

(i)   The bending moment diagram is zero when passing through a pin joint.
(ii) A pin joint does not alternate the variation of shear force.
(iii) An abrupt change of shear force occurs on the section where an intermediate support is applied.

EXAMPLE 4.3
The beam shown in Figure E4.3 is pinned to the wall at A. A vertical bracket BD is rigidly fixed to the beam at B, and a tie ED is pinned to the wall at E and to the bracket at D. The beam AC is subjected to a uniformly distributed load of 2 tons/ft and a concentrated load of 8 tons at C. Draw the shear force and bending moment diagrams for the beam.

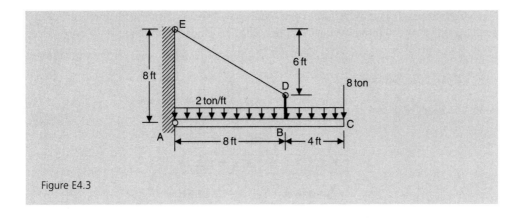

Figure E4.3

[Solution] *This is a statically determinate structure. The axial force in the tie can be resolved into vertical and horizontal components acting at D. The two force components generate a vertical force and a concentrated moment applied at the point B of the beam.*

Step 1: Calculate force and moment at B and reaction forces.

Resolving vertically:

$$R_A^V + F_D^V - 8\text{ton} - 2\text{tons/ft} \times 12\,\text{ft} = 0$$

Taking moment about A $(\widehat{A})$:

$$F_D^V \times 8\,\text{ft} + F_D^H \times 2\,\text{ft} - 8\,\text{tons} \times 12\,\text{ft} - 2\,\text{tons/ft} \times 12\,\text{ft} \times 6\,\text{ft} = 0$$

Due to the fact that the resultant of F_D^H and F_D^V must be in line with DE:

$$\frac{F_D^V}{F_D^H} = \frac{6\,\text{ft}}{8\,\text{ft}} = \frac{3}{4}$$

or

$$F_D^V = \frac{3}{4} F_D^H$$

Considering the above three equations, we have:

$$F_D^H = 30 \text{ tons } (\leftarrow)$$
$$F_D^V = 22.5 \text{ tons } (\uparrow)$$
$$R_A^V = 9.5 \text{ tons } (\uparrow)$$

The horizontal reaction at A, R_A^H, does not induce any shear or bending of the beam and is not included in the following calculation. The horizontal force, F_D^H, at D induces a concentrated moment applied at B. The moment caused by F_D^H is:

$$F_D^H \times 2\,ft = 30^T \times 2\,ft = 60\,tons.\,ft$$

Hence the beam is loaded as:

Step 2: Calculate shear forces and bending moments on critical cross-sections. The cross-sections at A , B and C are the critical sections.

(a) Section at A

$$M_A = 0$$
$$V_A = R_A^V = 9.5\,tons$$

(b) Section at B
Taking cut at the left-hand side of B:

Resolving vertically:

$$R_A^V - 2\,tons/ft \times 8\,ft - V_B = 0$$
$$V_B = R_A^V - 2\,ton/ft \times 8\,ft = -6.5\,tons\,(\uparrow)$$

Taking moment about B $(\overset{\frown}{B})$:

$$-R_A^V \times 8\,ft + 2\,tons/ft \times 8\,ft \times 4\,ft + M_B = 0$$
$$M_B = R_A^V \times 8\,ft - 2\,tons/ft \times 8\,ft \times 4\,ft = 12\,tons.\,ft$$

(c) Section at C

$$M_C = 0 \quad\text{and}$$
$$V_C = 8\,tons$$

Step 3: Draw the shear force and bending moment diagrams.
Due to the distributed load, the shear force diagrams between A and B, and B and C are both sloping lines, while the bending moment diagrams are parabolas. At B, due to the concentrated force and bending moment, abrupt changes occur in both diagrams.

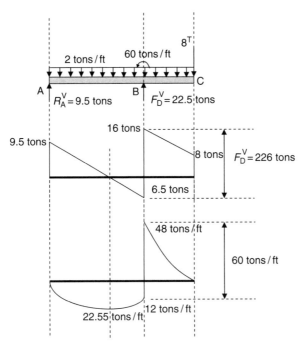

EXAMPLE 4.4
The simply supported beam shown in Figure E4.4 is loaded with the triangularly distributed pressure. Draw the shear force and bending moment diagrams of the beam.

Figure E4.4

[Solution] *This is statically determinate beam subjected to linearly distributed load. From Table 4.2, a uniformly distributed load produces a linear shear force and a parabolic bending moment diagram. From the relationship shown in Equations (4.1) and (4.2), the shear force and bending moment diagrams of this case are, respectively, parabolic and cubic.*

Step 1: Calculate support reactions.

Resolving vertically:

$$R_A + R_C = 2 \times \left(\frac{1}{2} \times 2\,m \times 10\,kN/m \right) = 20\,kN$$

Taking moment about A ($\stackrel{\frown}{A}$):

$$R_C \times 4 - 20\,kN \times 2\,m = 0$$

$$R_C = 10\,kN(\uparrow)$$

Thus:

$$R_A = 10\,kN(\uparrow)$$

Step 2: Calculate shear forces and bending moments on critical sections. Sections at A , B (change in distribution pattern) and C are the critical sections.

(a) *Section at A (Pin)*:

$$V_A = R_A = 10\,kN(\uparrow)$$

$$M_A = 0$$

(b) *Section at the immediate left of B*:

$$R_A = 10\,kN$$

Resolving vertically:

$$R_A - V_B - \frac{1}{2} \times 2\,m \times 10\,kN/m = 0$$

$$V_B = \frac{1}{2} \times 2\,m \times 10\,kN/m - 10\,kN = 0$$

Taking moment about B ($\stackrel{\frown}{B}$):

$$M_B - R_A \times 2\,m + \frac{1}{2} \times 2\,m \times 10\,kN/m \times \frac{2}{3} \times 2\,m = 0$$

$$M_B = 13.33\,kN\,m\left(\stackrel{\frown}{\,}\right)$$

(c) *At C (Roller)*:

$$V_C = -R_C = -10\,kN(\uparrow)$$

$$M_C = 0$$

Step 3: Draw the shear force and bending moment diagrams.
The shear force and bending moment diagrams between the critical sections can be sketched
as parabolas and cubes, respectively. If more accurate curves are required, shear forces and
bending moments on additional sections between each pair of the consecutive critical sections
are needed. For this purpose, it is appropriate to establish the general equations of shear force
and bending moment between these critical sections.

Between A and B: Take an arbitrary section between A and B, assume the distance from the
section to A is x and consider the equilibrium of the beam segment shown in the figure below.

Resolving vertically:

$$R_A - \frac{1}{2} \times x \times 5x - V(x) = 0$$

$$V(x) = 10 - \frac{5}{2}x^2 \,(kN)(\downarrow) \qquad (0 \le x \le 2\,m)$$

Taking moment about the arbitrary section $(\overset{\frown}{+})$:

$$M(x) - R_A \times x + \frac{1}{2} \times x \times 5x \times \frac{1}{3} \times x = 0$$

$$M(x) = 10x - \frac{5}{6}x^3 \qquad (0 \le x \le 2\,m)$$

Between B and C: Take an arbitrary section between B and C, assume the distance from the
section to C is x and consider the equilibrium of the beam segment shown in the following
figure.

Resolving vertically:

$$V(x) + R_C - \frac{1}{2} \times x \times 5x = 0$$

$$V(x) = -10 + \frac{5}{2}x^2 \,(kN) \qquad (0 \le x \le 2\,m)$$

Taking moment about the arbitrary section:

$$M(x) - R_C \times x + \frac{1}{2} \times x \times 5x \times \frac{1}{3} \times x = 0$$

$$M(x) = 10x - \frac{5}{6}x^3 \qquad (0 \le x \le 2\,\text{m})$$

The shear forces and bending moments on an arbitrary section can be calculated by introducing the x coordinate into the general equation derived above. For example,
At $x = 1\,\text{m}$ from A between A and B:

$$V(x) = 10 - \frac{5}{2}x^2 = 10 - \frac{5}{2} \times 1^2 = 7.5(\text{kN})(\downarrow)$$

$$M(x) = 10x - \frac{5}{6}x^3 = 10 \times 1 - \frac{5}{6} \times 1^3 = 9.17\,\text{kN m} \left(\begin{array}{c}\curvearrowright\end{array}\right)$$

At $x = 1.2\,\text{m}$ from C between B and C:

$$V(x) = -10 + \frac{5}{2}x^2 = -10 + \frac{5}{2} \times 1.2^2 = -6.4(\text{kN})(\downarrow)$$

$$M(x) = 10x - \frac{5}{6}x^3 = 10 \times 1.2 - \frac{5}{6} \times 1.2^3 = 10.56\,\text{kN/m} \left(\begin{array}{c}\curvearrowleft\end{array}\right)$$

Hence the shear force and bending moment diagrams are sketched as follows:

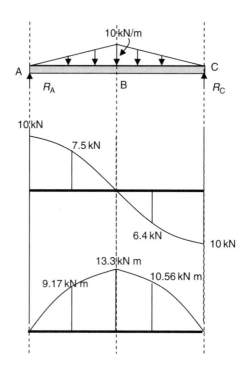

The following are observations from the above shear force and bending moment diagrams:

(i)   If a symmetric beam is subjected to symmetric loads, the shear stress is zero on the cross-section of symmetry and antisymmetric about the section.
(ii) If a symmetric beam is subjected to symmetric loads, the bending moment diagram is symmetric about the section of symmetry.
(iii) The distribution of shear force along the axis of a beam is one order higher than the order of the distributed transverse load.
(iv) The distribution of bending moment along the axis of a beam is two order higher than the order of the distributed transverse load.

EXAMPLE 4.5
The column—beam system shown in Figure E4.1 is subjected to a horizontal pressure q. Draw the shear force and bending moment diagrams.

Figure E4.1

[Solution] *This is a determinate frame system. The internal forces can be computed by following the general procedure described in Section 4.8 for single beams.*

Step 1: Replace supports by reactions.

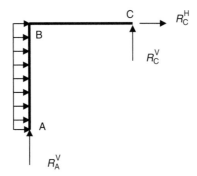

Due to the roller pin at A, the horizontal reaction at the support is zero.

Step 2: Derive solution of the reactions.
Take moment about C: ($\hat{C}$)

$$qa \times \frac{a}{2} - R_A^V \times a = 0$$

$$R_A^V = \frac{qa}{2}$$

Resolve vertically:

$$R_A^V + R_C^V = 0$$

$$R_C^V = -R_A^V = -\frac{qa}{2}$$

Step 3: Identify critical sections.
Sections at A, B and C are critical sections. In addition to the critical sections specified in Section 4.8, the section at which there is a sudden change in member orientation is also taken as a critical section.

Step 4: Calculate shear forces and bending moments on the critical sections.

(a) *Section at A*
   The section is supported by a roller pin. Since $R_A^H = 0$, there is no shear force. The bending moment at the support is also zero. Thus:

$$V_A = 0$$
$$M_A = 0$$

(b) *Section at B*

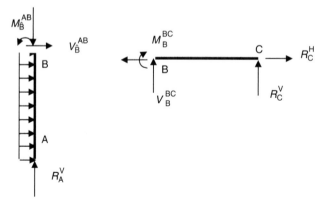

For the column, resolving horizontally:

$$V_B^{AB} + qa = 0$$

$$V_B^{AB} = -qa$$

Taking moment about ($\hat{B}$):

$$M_B^{AB} + qa\frac{a}{2} = 0$$

$$M_B^{AB} = -\frac{qa^2}{2}$$

For the beam, resolving vertically:

$$V_B^{BC} + R_C^V = 0$$
$$V_B^{BC} = -R_C^V = \frac{qa}{2}$$

Taking moment about $(\overset{\frown}{B})$:

$$a \times R_C^V - M_B^{BC} = 0$$
$$M_B^{BC} = -aR_C^V = -\frac{qa^2}{2}$$

(c) *Section at C*
The pin at C prevents the section from any vertical displacement. The vertical support reaction is the shear force. The bending moment is zero:

$$V_C = R_C^V = -\frac{qa}{2}$$
$$M_C = 0$$

Step 5: Draw the shear force and bending moment diagrams.

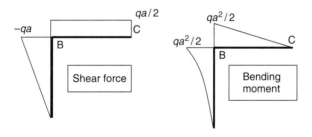

In the above diagrams, the sign of shear force is shown, while the bending moment diagram is drawn along the tension side of each member.

(i) The internal force diagram of a frame can be drawn by following exactly the same procedure followed for a single beam.
(ii) At a rigid joint of two members, the bending moment is constant across the joint if the there is no external moment applied at the location.
(iii) At a rigid joint of two members, the shear force is not necessarily constant across the joint even if there is no concentrated load applied at the location.

4.10 Conceptual questions

1. What are the general conditions for a beam to be in equilibrium?
2. What is a 'cantilever' and how is its equilibrium maintained?
3. What is a 'simply supported beam' and how is its equilibrium maintained?
4. Explain the terms 'shear force' and 'bending moment'.

5. Explain how the shear force and bending moment can be found on a section of a beam?
6. What are shear force and bending moment diagrams?
7. What are the relationships between shear force, bending moment and distributed load, and how are the relationships established?
8. A cantilever is loaded with a triangularly distributed pressure. Which of the following statements is correct?

(a) The shear force diagram is a horizontal line and the bending moment diagram is a sloping line.
(b) The shear force diagram is a sloping line and the bending moment diagram is a parabola.
(c) The shear force diagram is a parabola and the bending moment diagram is a cubic function.
(d) Both the shear force and the bending moment diagrams are parabolas.

9. A span of a beam carries only concentrated point loads, the shear force diagram is a series of _____ and the bending moment diagram is a series of _____.
10. The shear force diagram of a simply supported beam is shown in Figure Q4.10. Which of the following observations are not correct?

Figure Q4.10

(a) There is a concentrated force applied on the beam.
(b) There is no concentrated moment applied on the beam.
(c) There is a uniformly distributed load applied on the beam.
(d) There is a linearly distributed load applied on the beam.

11. The bending moment diagram of a simply supported beam is shown in Figure Q4.11. Identify the types of loads applied on the beam.

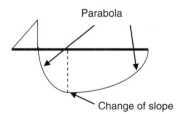

Parabola

Change of slope

Figure Q4.11

12. Consider a cantilever subjected to concentrated point loads. The maximum bending moment must occur at the fixed end (Yes/No).

13. At a point where the shear force diagram passes through zero the bending moment is either a maximum or a minimum (Yes/No).
14. Over any part of a beam where the shear force is zero, the bending moment has a constant value (Yes/No).
15. The cantilever is loaded as shown in Figure Q4.15. Which of the following four bending moment diagrams is correct?

Figure Q4.15

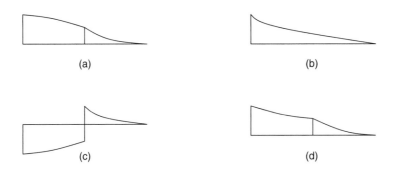

(a)   (b)

(c)   (d)

4.11 Mini test

Problem 4.1:   The shear force diagram over a segment of a beam is shown in the figure. Which of the following statements is not correct?

Parabola

Figure P4.1

(a) A concentrated point load is applied.
(b) A uniformly distributed load is applied.
(c) A concentrated moment may be applied.
(d) A triangularly distributed load is applied.
(e) No concentrated moment is applied.

Figure P4.2

Problem 4.2:   From the loaded beam and its shear force and bending moment diagrams shown in Figure P4.2, complete the following statements:

(a) The difference between the shear forces at c and d is _____.
(b) The slope of line a–b is _____.
(c) The difference between the bending moments at h and i is _____.
(d) The slopes of line i–j and line j–k are, respectively, _____ and _____.

Problem 4.3:   A simply supported beam is subjected to a total of load *P* that is distributed differently. Calculate maximum shear forces and maximum bending moments for each case and compare the results. What conclusion can you draw from the comparisons?

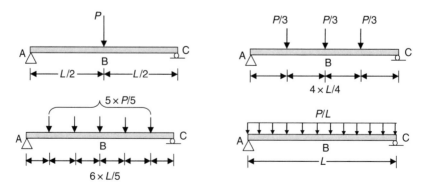

Figure P4.3

Problem 4.4:   For the beam loaded as shown in the figure draw the shear force and bending moment diagrams.

Figure P4.4

Problem 4.5:   Plot the shear force and bending moment diagrams of the beam loaded as shown in the figure. Sketch the curve of deflection of the beam.

Figure P4.5

5 Bending stresses in symmetric beams

In a beam subjected to transverse loads applied within the plane of symmetry (Figure 5.1(a)), only bending moment and shear forces develop on the cross-section (Figure 5.1(b)). The bending moment and shear force are, respectively, the resultants of the normal stresses (Figure 5.1(c)) and the shear stresses (Figure 5.1(d)) on the cross-section. The deformation of the beam is characterized by the following:

- The longitudinal fibres on the top side of the beam contract (are shortened).
- The longitudinal fibres on the bottom side of the beam extend (are elongated).
- Between the top and the bottom sides, there is a parallel surface within which fibres are neither contracted nor extended.
- Due to the uneven elongation of the fibres, the beam exhibits lateral (transverse) deformation that is termed as *deflection*.

Two important beam deformation terminologies (Figure 5.2) are introduced on the basis of the above conceptual analysis of beam deformation.

Figure 5.1

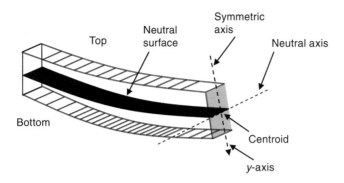

Figure 5.2

> - The surface formed by the fibres that are neither contracted nor extended is called *neutral surface*. This surface lies inside the beam between the top and the bottom surfaces. The beam fibres under the neutral surface are stretched and are in tension, while the beam fibres above the neutral surface are compressed and thus in compression.
> - The intersection of the neutral surface and a cross-section is called *neutral axis*. This intersection is a line within the cross-section and passes through the centroid of the section. On the cross-section, the area under the neutral axis is in tension, while the area above the neutral axis is in compression.

5.1 Normal stresses in beams

The normal stresses on a cross-section of a beam subjected to bending are calculated on the basis of the following basic assumptions:

> - A cross-sectional plane, taken normal to the beam's axis, remains plane throughout the deformation.
> - The strains in the beam's fibres vary linearly across the depth, proportional to their distances from the neutral axis.
> - Hooke's law is applicable to the individual fibres, that is, stress is proportional to strain.

The general expression for normal stresses caused by bending at a section is given as:

$$\sigma = \frac{My}{I} \qquad\qquad (5.1a)$$

where

  σ – normal stress at an arbitrary point on the section
  M – bending moment acting on the section
  y – distance from the neutral axis to the arbitrary point with positive y for points within the area in tension
  I – second moment of inertia of cross-section, representing a geometric quantity of the cross-section and having a unit of, for example, m^4. The mathematical expression of I is:

$$I = \int_A y^2 dA(\text{rea}) \qquad\qquad (5.1b)$$

The second moment of inertia of cross-section is an equivalent geometric quantity to the polar second moment of inertia, J, in torsion and the cross-sectional area, A, in axial tension and compression. The respective stresses are inversely proportional to the geometrical quantities of cross-sections.

The maximum normal stress occurs at the farthest fibres away from the neutral axis, that is, at $y = y_{max}$:

$$\sigma_{max} = \frac{My_{max}}{I} = \frac{M}{I/y_{max}} = \frac{M}{W} \tag{5.2}$$

where $W = I/y_{max}$ is called elastic modulus of section.

An efficient beam section has a small ratio of cross-sectional area and elastic modulus of section, A/W. Thus, in terms of the normal stress due to bending, an efficient section concentrates as much material as possible away from the neutral axis. This is why I-shaped sections are widely used in practice.

The elastic modulus of section reflects the geometric properties of a section and is different to the elastic modulus of materials, which depends only on material properties.

5.2   Calculation of second moment of inertia

The second moment of inertia is defined by the integral of $y^2 dA$ over the entire cross-sectional area with respect to an axis that is usually taken as the neutral axis. It is constant for a given section. The following procedure can be followed to compute the quality.

(a)   Find the centroid of the section.

- For a section having two axes of symmetry, the centroid lies at the intersection of the two axes.
- For a section having only one axis of symmetry the centroid lies on the axis. For the section shown in Figure 5.3, the y coordinate of the centroid, $\bar{y}$, is given by:

$$\bar{y} = \frac{\int_A y dA}{A} \tag{5.3a}$$

where $\int_A y dA$ is called *first moment of area* and A is the entire area of the section. From Equation (5.3a) it can be seen that when the area and the location of centroid of a section are known, the first moment of area can be easily computed by:

$$\int_A y dA = \bar{y} \times A \tag{5.3b}$$

Figure 5.3

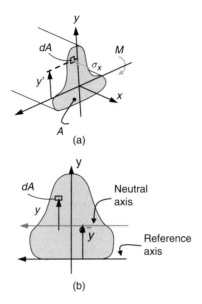

Figure 5.4

If A consists of a number of small sub-area(s) A_i and the y coordinates of their centroids are, respectively, $\bar{y}_i$, Equation (5.3b) is equivalent to:

$$\sum_i A_i \bar{y}_i = \bar{y} \times A \tag{5.3c}$$

(b) Find the neutral axis.

- For a section having two axes of symmetry, the neutral axis is one of them depending on the direction of bending moment applied on the section.
- For a section having single axis of symmetry, the neutral axis is perpendicular to the axis of symmetry and passes though the centroid of the section (Figure 5.4).

(c) Calculate second moment of inertia.

- The area integration is usually necessary only for a few regular shapes such as rectangles and circles. Most cross-sectional areas used in practice may be broken into an assembly of these regular shapes. Table 5.1 lists the second moments of inertia for some of the most commonly used shapes of section.
- For a complex cross-sectional area that is an assembly of the above regular shapes, the *parallel axis theorem* is used to compute the moment of inertia.

Table 5.1 Second moment of inertial about neutral axis

$I = \dfrac{1}{12}a^4$	$I = \dfrac{1}{12}bh^3$	$I = \dfrac{\pi}{64}D^4 = \dfrac{\pi}{4}R^4$	$I = \dfrac{\pi}{4}(R_{out}^4 - R_{in}^4)$

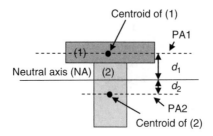

Figure 5.5

The parallel axis theorem states that the moment of inertia of an area, I, about an arbitrary axis equals the moment of inertia of the same area about a parallel axis passing through the area's centroid, I_0, plus the product of the area, A, and the square of the distance between the two axes, d, i.e.,

$$I = I_0 + Ad^2 \tag{5.4}$$

For example, the T section shown in Figure 5.5 is composed of two rectangles. The moment of inertia of the entire T section about its neutral axis is obtained by adding the moments of inertia of the two rectangles about the same axis, that is:

$$I_{NA}(\mathbf{T}) = I_{NA}^{(1)}(\blacksquare) + I_{NA}^{(2)}(\blacksquare)$$

where

$$I_{NA}^{(1)}(\blacksquare) = I_{PA1}^{(1)} + A^{(1)}(d_1)^2$$
$$I_{NA}^{(2)}(\blacksquare) = I_{PA2}^{(2)} + A^{(2)}(d_2)^2$$

$I_{PA1}^{(1)}$ = moment of inertia of area (1) about axis PA1
$I_{PA2}^{(2)}$ = moment of inertia of area (2) about axis PA2
$A^{(1)}$ = area of section (1)
$A^{(2)}$ = area of section (2)
PA1 = parallel axis passing through the centroid of $A^{(1)}$
PA2 = parallel axis passing through the centroid of $A^{(2)}$

5.3   Shear stresses in beams

In Figure 5.1(d), it can be noted the shear force on the section causes shear stress distribution. It can be conceptually argued that the vertical shear stress along the top and bottom boundaries of the section must vanish. This is because of the fact that shear stresses always occur in pair on two perpendicular planes (Figure 5.6). Since there is no shear stress on the top and bottom surfaces of the beam, the vertical shear stresses on the cross-section along the intersection of, for example, the top surface and the cross-section must also be zero. Thus, the distribution of shear stress across the depth of a beam, τ, can be parabolic (zero along the top and bottom sides and nonzero in between). For prismatic beams subjected to bending, the shear stress is

Figure 5.6

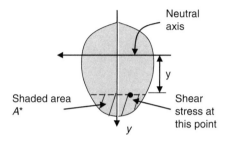

Figure 5.7

distributed parabolically, depending on the distance to the neutral axis. For symmetric bending, shear stresses are constant along any straight lines that are parallel to the neutral axis:

$$\tau = \frac{VS^*}{bI}$$ (5.5a)

where

V – the shear force acting on the section
b – the breadth of the beam at the location where the shear stress is computed
I – the second moment of inertia of the entire section about the neutral axis
S^* – the first moment of area about the neutral axis for the area enclosed by the boundary and the parallel line passing through the point at which shear stress is computed, that is, for the shaded area of Figure 5.7:

$$S^* = \int_{A^*} y \, dA^*$$ (5.5b)

5.4 Key points review

- The centroid is the center of a section.
- The moment of inertia of an area represents the geometric contribution of a cross-section to the bending resistance.
- The moment of inertia of an area about any axis is the moment of inertia of the area about a parallel axis passing through the area's centroid, plus its area times the square of the distance from the centroid to the axis (parallel axis theorem).
- A cross-sectional plane remains plane after bending deformation.
- There is no normal (axial) stress or strain due to bending at the neutral axis.
- Normal stress is compressive on one side of the neutral axis and tensile on the other side.
- The bending moment on a cross-section is the resultant of the normal stresses distributed on the same section.

- Normal strains and stresses are linearly distributed across the depth of a beam, proportional to the distance from the neutral axis.
- If a cross-section is made of one material, the normal stress is distributed continuously across the depth.
- Maximum normal stress occurs at a point furthest from the neural axis.
- The ratio between the second moment of inertia and the distance from neutral axis to the farthest point is termed *elastic modulus of section* and is a very important design parameter.
- An efficient section in bending concentrates more material away from the neutral axis such that a maximum elastic modulus of section (W) can be achieved with the use of a minimum cross-sectional area (A).
- For materials with different tensile and compressive strength, a shift of the neutral axis from the mid-depth of sections is desirable.
- Shear stress is distributed parabolically across the depth of beam.
- For a section with constant breadth, b, maximum shear stress occurs along the neutral axis.
- For a section with variable breadth, b, maximum shear stress occurs at the location where S^*/b is maximum.
- Shear stress distribution discontinues (abrupt increase or decrease) at the locations where the breadths of section have abrupt changes.
- The shear force on a cross-section is equal to the resultant of the shear stresses distributed on the same section.

5.5   Recommended procedure of solution

5.6   Examples

EXAMPLE 5.1

A beam of hollow circular section is loaded with concentrated point loads as shown in Figure E5.1. The inside and outside diameters of the hollow circular section are 45 mm and 60 mm, respectively. Calculate the maximum normal stress of the beam.

Figure E5.1(a)

[Solution] *For a beam with constant cross-section, the maximum normal stress and the maximum bending moment always occur on the same cross-section. A bending moment diagram is essential to determine the maximum bending moment and its location. This can be done by following the procedure described in Chapter 4.*

Step 1: Compute support reactions (Figure E5.1(b)).
Taking anticlockwise moment about A (↻)

Figure E5.1(b)

$$R_D \times 1.4\,m - 5\,kN \times 0.4\,m - 3\,kN \times 1.2\,m - 3\,kN \times 1.7\,m = 0$$

$$R_D = 7.64\,kN(\uparrow)$$

Resolving vertically:

$$R_A + R_D - 5\,kN - 3\,kN - 3\,kN = 0$$

$$R_A = 11\,kN - V_D = 3.36\,kN(\uparrow)$$

Step 2: Compute the bending moments on the critical sections at A, B, C, D and E (Figure E5.1(c)).

Figure E5.1(c)

(a) *Section at A*
   Due to the pin support:

$$M_A = 0$$

(b) *Section at B*
   Taking anticlockwise moment about B:

$$M_B - R_A \times 0.4\,m = 0$$

$$M_B = 3.36\,kN \times 0.4\,m = 1.34\,kN\,m \; (\curvearrowright)$$

(c) *Section at C*
   Taking anticlockwise moment about C:

$$M_C - R_A \times 1.2\,m + 5\,kN \times 0.8\,m = 0$$

$$M_C = 3.36\,kN \times 1.2\,m - 5\,kN \times 0.8\,m = 0.032\,kN\,m \; (\curvearrowright)$$

(d) *Section at D*
   Taking anticlockwise moment about D:

$$M_D - R_A \times 1.2\,m + 5\,kN \times 1.0\,m + 3\,kN \times 0.2\,m = 0$$

$$M_D = 3.36\,kN \times 1.4\,m - 5\,kN \times 1.0\,m - 3\,kN \times 0.2\,m$$

$$= -0.896\,kN\,m \; (\curvearrowleft)$$

(e) *Section at E*
    Due to the free end:

$$M_E = 0$$

Step 3: Draw the bending moment diagram (Figure 5.1(d)).
Since there is no distributed load applied between the critical sections, the bending moment distributions between these sections are slopping lines.

Figure E5.1(d)

It is clear from the bending moment diagram that the maximum magnitude of bending moment occurs on the section at B, that is:

$$M_{max} = 1.34 \, \text{KN m}$$

Thus, the maximum normal stress occurs on the same section.

Step 4: Compute second moment of inertia.
From Table 5.1 the moment of inertia of the hollow circular section:

$$I = \frac{\pi}{64}(R_{out}^4 - R_{in}^4) = \frac{\pi}{64}(D_{out}^4 - D_{in}^4) = \frac{\pi}{64}(60^4 - 45^4) \, \text{mm}^4$$
$$= 4.35 \times 10^{-7} \, \text{m}^4$$

Step 5: Compute normal stresses.
From Equation (5.1),
at B:

$$\sigma_{max} = \frac{M_{max} y_{max}}{I} = \frac{M_{max} D_{out}/2}{I} = \frac{1.34 \, \text{kN m} \times 30 \times 10^{-3} \, \text{m}}{4.35 \times 10^{-7} \, \text{m}^4}$$
$$= 92.41 \, \text{MN/m}^2$$

EXAMPLE 5.2
A simply supported steel beam is subjected to a uniformly distributed load as shown in Figure E5.2. The maximum allowable normal stress of the material is 160 MPa. Design the beam with a circular section, a rectangular section of $h/b = 2$ and an I-shaped section, respectively.

(Continued)

EXAMPLE 5.2   *(Continued)*

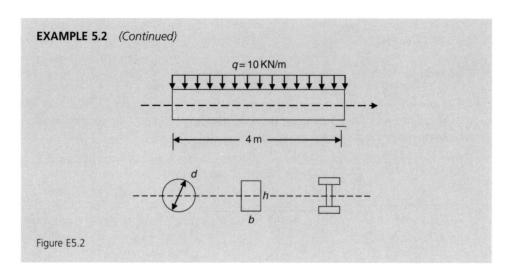

Figure E5.2

[Solution] *The maximum bending moment occurs at the mid-span. The maximum normal stress on the cross-section at the mid-span must be calculated first. The stress is then compared with the allowable stress of the material to determine the size of the cross-section.*

Step 1: Compute the reaction forces at the supports.
Since the beam is symmetrically loaded, the two support reactions are equal to half of the total applied load and act vertically upwards.

$$R = \frac{1}{2}(10 \times 4) = 20\,\text{kN}$$

Step 2: Compute the mid-span (maximum) bending moment:

$$M = 2R - 10 \times 2 \times 1 = 40 - 20 = 20\,\text{kN m}$$

Step 3: Section design.
The maximum normal stress on the section at the mid-span is:

$$\sigma = \frac{My_{max}}{I} = \frac{M}{I/y_{max}} = \frac{M}{W}$$

(a)   *For a circular section*:

$$I = \frac{\pi d^4}{64}$$

$$y_{max} = \frac{d}{2}$$

$$W = \frac{I}{y_{max}} = \frac{\pi d^3}{32}$$

Thus:

$$\sigma = \frac{M}{I/y_{max}} = \frac{20}{\pi d^3/32} \leq 160 \times 10^6$$

$$d \geq \sqrt[3]{\frac{20 \times 10^3 \times 32}{\pi \times 160 \times 10^6}} = 10.838 \times 10^{-2} \, m = 10.84 \, cm$$

The minimum area of the circular section:

$$A_c = \frac{\pi d^2}{4} = \frac{\pi \times 10.84^2}{4} = 92.29 \, cm^2$$

(b)   For a rectangular section ($h = 2b$):

$$I = \frac{1}{12}bh^3 = \frac{1}{12}b(2b)^3 = \frac{2}{3}b^4$$

$$y_{max} = \frac{h}{2} = b$$

Thus:

$$\sigma = \frac{M}{I/y_{max}} = \frac{20}{2b^3/3} \leq 160 \times 10^6$$

$$b \geq \sqrt[3]{\frac{3 \times 20 \times 10^3}{2 \times 160 \times 10^6}} = 5.724 \times 10^{-2} \, m = 5.72 \, cm$$

The minimum area of the rectangular section:

$$A_c = b \times h = 2b^2 = 2 \times 5.72^2 = 65.44 \, cm^2$$

(c)   For an I-shaped section (UB section):
From the British Standard (BS) sections, I/y_{max} is named as *elastic modulus of a section*, which is used as a design parameter:

$$\sigma = \frac{M}{I/y_{max}} = \frac{20}{I/y_{max}} \leq 160 \times 10^6$$

$$I/y_{max} \geq \frac{20 \times 10^3}{160 \times 10^6} = 0.125 \times 10^{-3} \, m^3 = 125 \, cm^3$$

The BS 5950 provides a table of elastic moduli sections for many UB (I-shaped) sections. The modulus that is immediately larger than 125 cm³ from the list is 153 cm³. The section designation of this modulus is UB 178 × 102 × 19 that has a cross-sectional area of 24.3 cm². Thus, the minimum area of the I-shaped section is:

$$A_I = 24.3 \, cm^2$$

The comparison for the weights of the beam using the three different sections can be made by comparing the cross-sectional areas of the sections. Obviously, amongst the three sections, the I-shaped section is the most economic one, which is more than three times lighter than the circular section.

EXAMPLE 5.3
A beam having the cross-section as shown in Figure E5.3 is subjected to a shear force of 12 kN and a bending moment of 12 kN m on a section. Compute (a) the normal stresses along the top and bottom surfaces of the hollow area; (b) the maximum magnitudes of the normal and shear stresses; and (c) the normal and shear stress distributions.

Figure E5.3(a)

[Solution] *The cross-section is not symmetric about any horizontal axis. Thus the vertical location of the centroid must be found first. The horizontal axis passing through the centroid is then the neutral axis. This can be done by treating the area enclosed by the outside boundary and deducting from it the hollow area enclosed by the inside boundary.*

Step 1: Compute the vertical distance of the centroid from the bottom side. This can be done in the following tabular form.

Area	A (mm^2)	$\bar{y}$(mm) from the bottom	$\int y \, dA = A\bar{y}$(mm^3)
Gross section	$160 \times 280 = 44,800$	140	$6,272,000$
Hollow area	$100 \times 80 = 8000$	$140 + 50 = 190$	$1,520,000$

Thus from Equation (5.3) for the actual section:

$$\sum A = 44,800 - 8000 = 36,800 \, \text{mm}^2$$

$$\sum A\bar{y} = 6,272,000 - 1520,000 = 4,752,000 \, \text{mm}^3$$

$$y_c = \frac{\sum A\bar{y}}{\sum A} = \frac{4,752,000}{36,800} = 129.1 \, \text{mm}$$

The neutral axis is 129.1 mm above and parallel to the bottom side of the section.

Step 2: Compute the moment of inertia from the parallel axis theorem.
For the gross area enclosed by the outside boundary (Figure E5.3(b)) the moment of inertia about the neutral axis is computed from Equation (5.4) as:

Figure E5.3(b)

$$I_{gross} = I_0 + Ad^2 = \frac{160 \times 280^3}{12}$$
$$+ 280 \times 160 \times (140 - 129.1)^2$$
$$= 2.98 \times 10^8 \, mm^4$$

For the hollow area (Figure E5.3(c)) the moment of inertia about the neutral axis is also computed from Equation (5.4) as:

Figure E5.3(c)

$$I_{hollow} = I_{out} + Ad^2 = \frac{90 \times 100^3}{12}$$
$$+ 100 \times 80 \times (190 - 129.1)^2$$
$$= 3.72 \times 10^7 \, mm^4$$

The moment of inertia of the cross-sectional area:

$$I = I_{gross} - I_{hollow}$$
$$= (29.8 - 3.72) \times 10^7 = 26.1 \times 10^7 \, mm^4$$

Step 3: Compute normal stresses.
Along the top surface of the hollow area
The distance from the neutral axis to the surface, that is, the y coordinate of the surface:

$$y = -(140 - 129.1 + 100)\,\text{mm} = -110.9\,\text{mm}$$

$$\sigma = \frac{My}{I} = \frac{12\,\text{kN m} \times (-110.9) \times 10^{-3}\,\text{m}}{26.1 \times 10^{-5}\,\text{m}^4} = -5.1\,\text{MN/m}^2 \text{ (compression)}$$

Along the bottom surface of the hollow area
The distance from the neutral axis to the surface, that is, the y coordinate of the surface:

$$y = -(140 - 129.1)\,\text{mm} = -10.9\,\text{mm}$$

$$\sigma = \frac{My}{I} = \frac{12\,\text{kN m} \times (-10.9) \times 10^{-3}\,\text{m}}{26.1 \times 10^{-5}\,\text{m}^4} = -0.5\,\text{MN/m}^2 \text{ (compression)}$$

Step 4: Compute the maximum magnitude of normal stress.
It is obvious from Equation (5.1) that the maximum magnitude of normal stress occurs along the top side of the section since its distance to the neutral axis is the maximum. This distance is:

$$y = -(280 - 129.1)\,\text{mm}$$

$$\sigma = \frac{My}{I} = \frac{12\,\text{kN m} \times (-150.9) \times 10^{-3}\,\text{m}}{26.1 \times 10^{-5}\,\text{m}^4} = -6.94\,\text{MN/m}^2 \text{ (compression)}$$

Step 5: Compute the maximum magnitude of shear stress.
From Equation (5.5) the maximum magnitude of shear stress may occur at the location where (a) the S^* defined in Equation (5.5) is maximum or (b) the breadth of the cross-section, b, is minimum.

(a) S^* is always maximum along the neutral axis. Consider the area below the neutral axis:

$$S^* = \sum A\bar{y} = 160\,\text{mm} \times 129.1\,\text{mm} \times \frac{129.1\,\text{mm}}{2} = 1.33 \times 10^{-3}\,\text{m}^3$$

$$\tau = \frac{VS^*}{bI} = \frac{12\,\text{kN} \times 1.33 \times 10^{-3}\,\text{m}^3}{160 \times 10^{-3}\,\text{m} \times 26.1 \times 10^{-5}\,\text{m}^4} = 0.382\,\text{MN/m}^2$$

(b) b is minimum between the top and the bottom surfaces of the hollow area, while with this range the S^* taken for the area immediately above the bottom side of the hollow part is the largest. Thus at this location (140 mm from the top):

$$S^* = \sum A\bar{y} = 140 \times 160 \times (140 - 129.1 + 70) - 100 \times 80 \times (140 - 129.1 + 50)$$

$$= 1.32 \times 10^{-3}\,\text{m}^3$$

$$\tau = \frac{VS^*}{bI} = \frac{12\,\text{kN} \times 1.32 \times 10^{-3}\,\text{m}^3}{2 \times 40 \times 10^{-3}\,\text{m} \times 26.1 \times 10^{-5}\,\text{m}^4} = 0.76\,\text{MN/m}^2$$

Therefore the maximum magnitude of shear stress occurs at the location 140 mm below the top surface of the cross-section. At the same location, if we take the breadth of the solid part as b in Equation (5.5), that is:

$$\tau = \frac{VS^*}{bI} = \frac{12\,\text{kN} \times 1.32 \times 10^{-3}\,\text{m}^3}{160 \times 10^{-3}\,\text{m} \times 26.1 \times 10^{-5}\,\text{m}^4} = 0.379\,\text{MN/m}^2$$

Observation: *Due to the abrupt change of the breadth of section when across the bottom side of the hollow area (from 160 mm to 80 mm), the shear stress jumps from 0.379 MN/m² to 0.76 MN/m². It can be concluded that on a cross-section if the breadth of section changes suddenly across a line, the shear stress also has a sudden change (increased or decreased) by the ratio of change in breadth, that is, the ratio of the breadth immediately below and above the line. In this example, the ratio is 160/80 = 2. The shear stress above the line (140 mm from the top) is therefore twice as big as it is below the line.*

By calculating the normal stress at the bottom surface and the shear stresses above and below the top side of the hollow area, the normal and shear stress distribution on the cross-section can be sketched as below:

Obviously, a sudden change in geometry of the cross-section does not introduce sudden change in normal stress, but in shear stress distribution.

EXAMPLE 5.4
Determine the maximum normal stresses in the concrete and the steel in a simply supported reinforced concrete beam subjected to uniformly distributed load of 1250 lb/ft over a span of 25 ft. The cross-section of the beam, as shown in Figure E5.4, is reinforced with three steel bars having a total cross-sectional area of π in². Assume that the ratio of E for steel to that for concrete is 15 and ignore all concrete in tension.

Figure E5.4

[Solution] *Since the stress-carrying capacity of concrete in tension zone is significantly smaller compared with steel, all concrete below the neutral axis is neglected. The bending moment on a cross-section of the beam is balanced by the compression in the concrete and the tension in the steel reinforcement. To compute the location of the neutral axis and stress distribution on the cross-section, the area of the steel needs to be transferred to its equivalent area of concrete on the basis of the ratio of E, so that all the formulas developed for beam-bending problems are valid. The transformation is based on the assumption that the strains and the resultant force in the steel reinforcement and the equivalent concrete are equal.*

Step 1: Compute support reactions.
Since the beam is symmetrically loaded, the two support reactions are both equal to half of the total applied load and act vertically upwards:

$$R = \frac{1250 \times 25}{2}\, \text{lb}$$

Step 2: Compute bending moment.
The maximum normal stress occurs on the cross-section subjected to the maximum bending. It is obvious that the maximum bending moment occurs at the mid-span. The bending moment at the mid-span is calculated as follows:
    Taking moment about the mid-span:

$$M = \frac{1250 \times 25}{2} \times 12.5 - 1250 \times 12.5 \times \frac{12.5}{2}$$

$$= 97656.25\, \text{lb/ft}$$

$$= 1171875\, \text{lb/in}$$

Step 3: Compute the equivalent concrete area A_{eq} of the steel.
Assume that on the equivalent concrete A_{eq} and the steel bars A_{steel}, both the strains and the resistance forces are the same. Thus:

$$\sigma_{steel} A_{steel} = \sigma_{eq} A_{eq} \tag{E5.3.1}$$

or

$$E_{steel} \varepsilon_{steel} A_{steel} = E_c \varepsilon_{eq} A_{eq}$$

Since:

$$\varepsilon_{steel} = \varepsilon_{eq}$$

$$A_{eq} = \frac{E_{steel}}{E_c} A_{steel} = 15 A_{steel} = 15\pi = 47.12\, \text{in.}^2 \tag{E5.3.2}$$

Thus, the steel reinforcement can be replaced by an equivalent concrete area of 47.12 in^2 (see Figure E5.4).

Step 4: Compute vertical distance of the centroid from the top.

Area	$A(\text{in}^2)$	$\bar{y}(\text{in})$ from the top	$\int y\,dA = A\bar{y}(\text{in}^3)$
Concrete in compression	$15 \times y_c$	$y_c/2$	$7.5y_c^2$
Equivalent concrete	47.12	25	1178

From Equation (5.3):

$$\sum A = (15y_c + 47.12)\,\text{in.}^2$$

$$\sum A\bar{y} = (7.5y_c^2 + 1178)\,\text{in.}^3$$

$$y_c = \frac{\sum A\bar{y}}{\sum A} = \frac{7.5y_c^2 + 1178}{15y_c + 47.12}$$

Hence:

$$7.5y_c^2 + 47.12y_c - 1178 = 0$$

$$y_c = 9.4\,\text{in.}$$

The neutral axis is 9.4 in. below the top surface.

Step 5: Compute moment of inertia of the transferred cross-section.
For the concrete (Figure E5.4):

$$I_c = \frac{1}{12} \times 15 \times y_c^3 + 15 \times y_c \times \left(\frac{y_c}{2}\right)^2$$

$$= \frac{1}{12} \times 15 \times 9.4^3 + 15 \times 9.4 \times \left(\frac{9.4}{2}\right)^2$$

$$= 4152.92\,\text{in.}^4$$

For the equivalent concrete area (Figure E5.4):

$$I_{eq} = 0 + A_{eq} \times (25 - y_c)^2 = 11467.12\,\text{in.}^4$$

For the entire section:

$$I = I_c + I_{eq} = 15620.04\,\text{in.}^4$$

Note that zero is given to the moment of inertia of the equivalent area of reinforcement about the parallel axis passing through its own centriod. This is because the thickness of the equivalent concrete area is very small and the resulting moment of inertia is far smaller than the second term in I_{eq}.

Step 6: Compute normal stresses.
Along the top the maximum compressive stress in concrete occurs:

$$\sigma = \frac{My}{I} = \frac{1171875 \times 9.4}{15620.04} = 705.2\,\text{lb/in. (compression)}$$

On the equivalent area of reinforcement the maximum tensile stress occurs:

$$\sigma_{eq} = \frac{My}{I} = \frac{1171875 \times (25 - 9.4)}{15620.04} = 1170.4 \, \text{lb/in. (tension)}$$

The actual tensile stress in the steel can be computed from Equations (E5.3.1) and (E5.3.2), that is:

$$\sigma_{steel} = \frac{A_{eq}}{A_{steel}}\sigma_{eq} = \frac{E_{steel}}{E_c}\sigma_{eq} = 15 \times 1170.4 = 17555.6 \, \text{lb/in}^2$$

EXAMPLE 5.5

Derive an expression for the shear stress distribution on a rectangular section subjected to a shear force V (Figure E5.5(a)).

Figure E5.5(a)

[Solution] *This question asks for a straightforward application of Equation (5.5). In order to find the expression of shear stress distribution, shear stress at an arbitrary location must be calculated. The cross-section is symmetric and the horizontal axis of symmetry is the neutral axis of the section.*

Step 1: Select an arbitrary point on the cross-section. The vertical distance from the point to the neutral axis is y ($0 \le y \le h/2$) (Figure E5.5(b)).

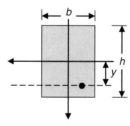

Figure E5.5(b)

Step 2: Draw a line passing through the selected point and parallel to the neutral axis. (Figure E5.5(b)).

Step 3: Calculate the first moment of area S^* (Equation (5.5)). The parallel line separates the cross-section into two parts, one of which *does not* include the neutral axis (Figure E5.5(c)). Usually the first moment of area of this part about the neutral axis is calculated as S^*.

Figure E5.5(c)

The area of the part:

$$A^* = b \times (\frac{h}{2} - y)$$

The distance from the neutral axis to the centriod of the part:

$$\bar{y} = y + \frac{1}{2}(\frac{h}{2} - y) = \frac{h}{4} + \frac{y}{2}$$

The first moment of area of the part about the neutral axis:

$$S^* = A^*\bar{y} = b(\frac{h}{2} - y) \times (\frac{h}{4} + \frac{y}{2})$$

$$= \frac{b}{2}[(\frac{h^2}{4} - y^2)]$$

Step 4: Compute shear stress distribution.
The moment of inertia of the section about its neutral axis is:

$$I = \frac{1}{12}bh^3$$

and the breadth of the section at the arbitrary location is *b*. Thus,

$$\tau = \frac{VS^*}{bI} = \frac{12V}{b(bh^3)} \times \frac{b}{2}[(\frac{h^2}{4} - y^2)]$$

$$= \frac{6V}{bh^3}[(\frac{h^2}{4} - y^2)]$$

The distribution of the shear stress is a parabolic function of the distance, *y*, from the neutral axis as shown in Figure E5.5(d). It is obvious that the maximum shear stress occurs along the neutral axis, that is, at $y = 0$.

Figure E5.5(d)

From the expression of distribution:

$$\tau_{max} = \frac{6V}{bh^3}[(\frac{h^2}{4} - y^2)]\Big|_{y=0} = \frac{3V}{2bh} = \frac{3V}{2A}$$

where $A(= bh)$ is the cross-sectional area of the beam. V/A represents the average shear stress on the section.

- *The maximum shear stress due to bending of a rectangular section is 1.5 times larger than the average shear stress on the section and occurs along the neutral axis.*

EXAMPLE 5.6
The compound beam shown in Figure E5.6(a) is composed of two identical beams of rectangular section: (a) if the two beams are simply placed together with frictionless contact and the maximum allowable normal stress of the material is $[\sigma]$, calculate the maximum value of the force, P, that can be applied on the beam; (b) if the two beams are fastened together by a bolt as shown in Figure E5.6(b), what is the maximum value of P; and (c) if the maximum allowable shear stress in the bolt is $[\tau]$, calculate the minimum diameter of the bolt when the compound beam is loaded with the P calculated from (b).

Figure E5.6

[Solution] *This example tests your understanding of bending deformation in relation to neutral axis. In Case (a), due to the frictionless contact, the two beams slide on each other and deform independently, each of which carries a half of the total bending moment induced by P. Therefore, the two beams have their own neutral axes. In Case (b), the two beams are fastened together so that the compound beam acts as a unit and has only one neutral axis.*

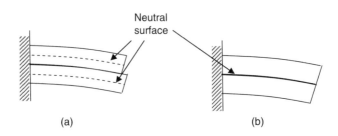

(a) The maximum normal stress occurs at the fixed end where the bending moment is the maximum:

$$M_{max} = PL$$

The maximum bending moment carried by each of the beams is therefore $PL/2$. Thus, by Equation (5.1), the maximum normal stress within each of the beams is:

$$\sigma = \frac{My}{I} = \frac{\frac{PL}{2} \times \frac{h}{4}}{\frac{1}{12} \times b \times \left(\frac{h}{2}\right)^3} = \frac{12PL}{bh^2} \leq [\sigma]$$

So the maximum force that can be applied on the beam is:

$$P_{max} \leq \frac{[\sigma]\,bh^2}{12L}$$

(b) When the two beams act as a unit, the contact surface is now the neutral surface. The maximum normal stress is then:

$$\sigma = \frac{My}{I} = \frac{PL \times \frac{h}{2}}{\frac{1}{12} \times b \times h^3} = \frac{6PL}{bh^2} \leq [\sigma]$$

So the maximum force that can be applied on the beam for this case is:

$$P_{max} \leq \frac{[\sigma]\,bh^2}{6L}$$

It can be seen that the bonded beam has higher load-carrying capacity than that of the two separated beams.

(c) Assume that the two beams are bonded perfectly together, and the maximum shear stress occurs along the neutral axis. Thus by Equation (5.4):

$$\tau_{max} = \frac{VS^*}{bI} = \frac{P_{max}}{b \times \frac{1}{12}bh^3} b \times \left(\frac{h}{2}\right) \times \left(\frac{h}{4}\right)$$

$$= \frac{[\sigma]\,bh^2}{6L} \frac{3}{2bh} = \frac{[\sigma]\,h}{4L}$$

Since the beam has a constant shear force distribution along the axis, this maximum shear stress applies to any cross-section of the beam. Thus the resultant of the shear stresses acting on the neutral surface is:

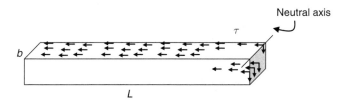

$$Q = \tau_{max} \times b \times L$$

In case (c), the shear force is entirely carried by the bolt that has a cross-sectional area of $\pi d^2/4$, where d is the diameter of the bolt. The shear stress in the bolt is:

$$\tau = \frac{Q}{A} = \frac{\tau_{max} bL}{\pi d^2/4} = \frac{[\sigma] h}{4L} \times \frac{bL}{\pi d^2/4} = \frac{[\sigma] hb}{\pi d^2} \le [\tau]$$

So the minimum diameter of the bolt is:

$$d \ge \sqrt{\frac{[\sigma] bh}{\pi [\tau]}}$$

5.7 Conceptual questions

1. What is meant by 'neutral axis'?
2. What is meant by 'the second moment of area of a cross-section?' If this quantity is increased, what is the consequence?
3. What is the parallel axis theorem, and when can it be used?
4. What is meant by 'elastic modulus of section?' Explain how it is used in steel design.
5. How is the normal stress due to bending distributed on a beam section?
6. Why do normal stresses due to bending vary across a beam's cross-section?
7. When a beam is under bending, the magnitudes of the maximum compressive and the maximum tensile stresses are always the same? (Y/N)
8. The cantilever shown in Figure 5.8 has an I-shaped section and is loaded with a uniformly distributed pressure. The dashed line passes through the centroid of the cross-section.

Figure Q5.8

Which of the following statements is correct?

(a) The dashed line is the neutral axis and the maximum tensile stress occurs in the fibre along the bottom surface on the cross-section at the fixed end.
(b) The dashed line is the neutral axis and the maximum tensile stress occurs in the fibre along the top surface on the cross-section at the fixed end.
(c) The solid line is the neutral axis and the maximum tensile stress occurs in the fibre along the bottom surface on the cross-section at the fixed end.
(d) The solid line is the neutral axis and the maximum tensile stress occurs in the fibre along the bottom surface on the cross-section at the fixed end.

9. Explain why symmetric sections about neutral axis are preferable for beams made of materials with equal tensile and compressive strengths, while unsymmetrical sections

about neutral axis are preferable for beams made of materials with different tensile and compressive strengths.

10.   When a beam is under bending, the maximum shear stress always occurs along the neutral axis? (Y/N)

11.   The cross-section shown in Figure Q5.11 is loaded with a shear force V. Which one of the following statements is correct when Equation (5.5) is used to calculate the shear stress along m–m.

Figure Q5.11

(a)   In Equation (5.5), S^* denotes the first moment of area of the entire cross-sectional area about the neutral axis and b is the width of the section along m–m.

(b)   In Equation (5.5), S^* denotes the first moment of area of the entire cross-sectional area about the neutral axis and b is the width of the section along the neutral axis.

(c)   In Equation (5.5), S^* denotes the first moment of area of the section below m–m about the neutral axis and b is the width of the section along m–m.

(d)   In Equation (5.5), S^* denotes the first moment of area of the section below m–m about the neutral axis and b is the width of the section along the neutral axis.

12.   Two beams of the same material and cross-section are glued together to form a combined section as shown in Figure Q5.12.

Figure Q5.12

If the beam is subjected to bending, which form of the following normal stress distributions is correct?

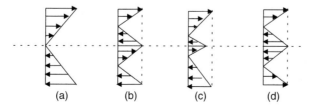

(a)   (b)   (c)   (d)

If the two beams are simply placed together with frictionless contact, which form of the above normal stress distributions is correct?

13. In Question 12, if the cross-sections of the two beams are channel-shaped, as shown in Figure Q5.13, and the contact between the two beams are frictionless, which form of the normal stress distributions shown in Question 12 is correct?

Figure Q5.13

14. The beam shown in Figure Q5.14 is composed of two identical planks that are placed either horizontally or vertically and not fastened together. If the beam is under pure bending, which one of the following statements is correct?

section (a)   section (b)

Figure Q5.14

(a) The maximum normal stresses in sections (a) and (b) are the same.
(b) The maximum normal stress in section (a) is greater.
(c) The maximum normal stress in section (b) is greater.
(d) Any of the above statements can be correct, depending on the magnitude of the bending moment applied.

15. On the cross-section of a steel beam under bending, which one of the following statements is correct?

(a) At the point where the maximum normal stress occurs, shear stress is always zero, while at the point where the maximum shear stress occurs, normal stress is not necessarily zero.
(b) At the point where the maximum normal stress occurs, shear stress is not zero, while at the point where the maximum shear stress occurs, normal stress is always zero.
(c) At the point where the maximum normal stress occurs, shear stress is always zero, and at the point where the maximum shear stress occurs, normal stress is also zero.
(d) At the point where the maximum normal stress occurs, shear stress is not zero, and at the point where the maximum shear stress occurs, normal stress is also not zero.

5.8   Mini test

Problem 5.1:   The following four cross-sections have the same cross-sectional area and are made of the same material. Which one of them is the most efficient section and which of them is the least efficient? Explain why.

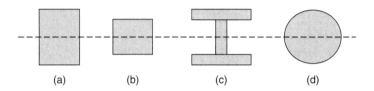

(a)   (b)   (c)   (d)

Figure P5.1

Problem 5.2:   Sketch shear stress distributions on the following cross-sections subjected to a vertical shear force.

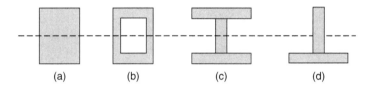

(a)   (b)   (c)   (d)

Figure P5.2

Problem 5.3:   A T beam shown in the Figure P5.2 is subjected to a uniformly distributed load. Determine the location of the cross-section on which maximum tensile stress and maximum compressive stress occur and calculate the magnitudes of these stresses. Plot the distribution of the normal stress on the cross-section.

Figure P5.3

Problem 5.4:   For the beam shown in Problem 5.3, Determine the location of the cross-section on which maximum shear stress occurs. Find an expression for the shear stress distribution on the section and calculate the magnitude of the maximum shear stress.

Problem 5.5: Determine the maximum allowable bending moment that can be applied to the composite beam section shown in Figure P5.4. The wood beam is longitudinally reinforced with steel strips on both the top and the bottom sides. The two materials are fastened together so that they act as a unit. The maximum allowable normal stresses of wood and steel are, respectively, 8.3 MN/m² and 140 MN/m². (E_{steel} = 200 GPa, E_{wood} = 8.3 GPa).

Figure P5.4

6 Deflection of beams under bending

A beam is any long structural member on which loads act perpendicular to the longitudinal axis. If the cross-section of the beam has a plane of symmetry and the transversely applied loads are applied within the plane, the axis of the beam will deflect from its original position within the plane of symmetry, as shown in Figure 6.1.

The shape of the deflection curve will depend on several factors, including:

- the material properties of the beam as measured by the elastic modulus of material;
- the beam's cross-section as measured by its second moment of inertia;
- the load on the beam, described as a function of the position along the beam; and
- the supports of the beam.

If the beam is placed in the x–y coordinate system as shown in Figure 6.1, the vertical deflection of the axis, y, is a function of the x coordinate and $\theta(x)$ denotes the slope of the deflection curve at an arbitrary x coordinate. Bending of a beam is measured by curvature of the deflected axis that can be approximately calculated by d^2y/dx^2. The deflection of a beam is characterized by the following:

- Deflection (positive downwards) $= y(x)$.
- Slope of deflection curve $= \dfrac{dy(x)}{dx}$.
- Curvature of deflection curve $= \dfrac{d^2y(x)}{dx^2}$.
- Flexural rigidity $= EI$, representing the stiffness of a beam against deflection. For a given bending moment and at a given section, a stiffer (greater) EI results in a smaller curvature.

Figure 6.1

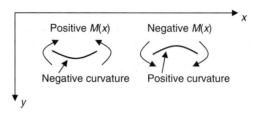

Figure 6.2

6.1 Sign convention

A positive bending moment is defined as a moment that induces *sagging* deflection. As shown in Figure 6.2, a positive moment always produces a negative curvature in the adopted coordinate system, where sagging deflection is defined as positive.

6.2 Equation of beam deflection

The extent that a beam is bent, that is, the curvature of deflection curve, is directly proportional to the applied bending moment $M(x)$, and is inversely proportional to the flexural rigidity EI:

$$\frac{d^2y(x)}{dx^2} = -\frac{M(x)}{EI} \tag{6.1}$$

For a beam with a uniform cross-section, from Equations (4.1) and (4.2):

$$\frac{d^3y(x)}{dx^3} = -\frac{1}{EI}\frac{dM(x)}{dx} = -\frac{V(x)}{EI} \tag{6.2}$$

$$\frac{d^4y(x)}{dx^4} = -\frac{1}{EI}\frac{dV(x)}{dx} = \frac{q(x)}{EI} \tag{6.3}$$

The negative sign on the right-hand side of Equation (6.1) reflects the sign difference between bending moment and the resulting curvature, following the sign convention described in Section 6.1. The deflection, $y(x)$, can be solved from the above differential equations by various methods of solution.

6.2.1 The integration method

The integration method requires integrating the differential equations of beam deflection up to four times, depending on whether Equation (6.1), or (6.2) or (6.3) is used. The general procedure of the integration method is as follows:

- Establish the equation of beam deflection in the form of either Equations (6.1) or (6.2) or (6.3) as appropriate. The choice of the above differential equations depends on whether or not an expression of either bending moment, or shear force, or applied distributed load can be formulated.
- Integrate twice, three times and four times when Equations (6.1), (6.2) and (6.3) are, respectively, used.
- Apply continuity conditions at the critical sections where either shear force or bending moment or applied load changes patterns of distribution. This means that at any

junction of two parts of a beam the deflection and the slope of the deflection curve must be the same unless they are jointed by a pin.
- Apply support (boundary) conditions for the solution of the integral constants introduced in Step 2.

The most commonly seen support conditions are listed in Table 4.1. Table 6.1 below lists some typical continuity conditions.

6.2.2 The superposition method

The superposition method can be used to obtain deflection of a beam subjected to multiple loads, particularly when the deflections of the beam for all or part of the individual loads are known from previous calculations or design tables, etc. For example, the beam shown in Figure 6.3(a) can be separated into three different cases. The algebraic sum of the three separate deflections caused by the separate loads gives the total deflection (Figure 6.3(b)).

Table 6.2 presents beam deflections for a number of typical load-support conditions that can be used as the separate cases to form the total solution of a complex problem.

The general procedure of the superposition method is as follows:

- Resolve a complex problem into several simpler problems whose deflections are readily available.
- Express the deflections of the simpler problems in a common coordinate system.
- Superimpose the deflections algebraically to compute the total deflection.

Table 6.1 Continuity conditions at critical sections

Type of critical section		Displacements	Internal forces
Intermediate roller support		Deflection $y_a = y_b = 0$ Slope $\dfrac{dy_a}{dx} = \dfrac{dy_b}{dx}$	Bending moment $M_a = M_b$
Intermediate pin		Deflection $y_a = y_b = 0$	Shear force $V_a = V_b$ Bending moment $M_a = M_b = 0$
Concentrated force		Deflection $y_a = y_b$ Slope $\dfrac{dy_a}{dx} = \dfrac{dy_b}{dx}$	Shear force $V_a - V_b = P$ Bending moment $M_a = M_b$
Concentrated moment		Deflection $y_a = y_b$ Slope $\dfrac{dy_a}{dx} = \dfrac{dy_b}{dx}$	Shear force $V_a = V_b$ Bending moment $M_a - M_b = M$
Discontinuity in distributed load		Deflection $y_a = y_b$ Slope $\dfrac{dy_a}{dx} = \dfrac{dy_b}{dx}$	Shear force $V_a = V_b$ Bending moment $M_a = M_b$

Figure 6.3

6.2.3 *Macaulay's method (step function method)*

A step function is denoted by the brackets "< >." For a variable $x - a$, we define:

$$\langle x - a \rangle^n = \begin{cases} (x-a)^n & x \geq a \\ 0 & x < a \end{cases} \tag{6.4}$$

where n is any real number. This function enables us to write the bending moment for the beam shown in Figure 6.4 in a single equation.

Table 6.2 Deflection of beam

Load and support	Deflection	Maximum deflection	End rotation
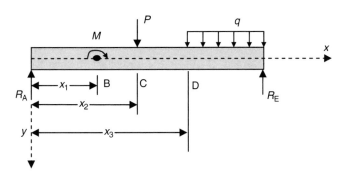	$y = \dfrac{Px^2}{6EI}(3a - x),$ $0 \le x \le a$ $y = \dfrac{Pa^2}{6EI}(3x - a),$ $a \le x \le l$	$y_{max} = y_B$ $= \dfrac{Fa^2}{6EI}(3l - a)$	$\theta_B = \dfrac{Fa^2}{2EI}$
	$y = \dfrac{qx^2}{24EI}(x^2 + 6l^2 - 4lx)$	$y_{max} = y_B$ $= \dfrac{ql^4}{8EI}$	$\theta_B = \dfrac{ql^3}{6EI}$
	$y = \dfrac{Pbx}{6lEI}(l^2 - b^2 - x^2)$ $0 \le x \le a$ $y = \dfrac{P}{EI}[\dfrac{bx}{6l}(l^2 - b^2 - x^2)$ $+ \dfrac{1}{6}(x - a)^3]$ $a \le x \le l$	$y_{max} = \dfrac{Pb}{9\sqrt{3}EIl}$ $\times \sqrt{(l^2 - b^2)^3}$ At $x = \dfrac{\sqrt{l^2 - b^2}}{3}$	$\theta_A = \dfrac{Pab(l + b)}{6lEI}$ $\theta_B = \dfrac{-Pab(l + a)}{6lEI}$
	$y = \dfrac{qx}{24EI}(l^2 - 2lx^2 + x^3)$	$y_{max} = \dfrac{5ql^4}{384EI}$	$\theta_A = -\theta_B$ $= \dfrac{ql^3}{24EI}$

Figure 6.4

The bending moments within different parts of the beam are as follows:

- Moment on sections within (A, B) $0 \le x \le x_1$ R_Ax
- Moment on sections within (B, C) $x_2 \le x \le x_2$ $R_Ax + M$
- Moment on sections within (C, D) $x_2 \le x \le x_3$ $R_Ax + M - P(x - x_2)$
- Moment on sections within (D, E) $x \ge x_3$ $R_Ax + M - P(x - x_2) - \frac{1}{2}q(x - x_3)^2$

By using the step function described in Equation (6.4), the bending moment on an arbitrary cross-section between A and E, which is x away from the left end of the beam, is:

$$M(x) = R_A x + M\langle x - x_1 \rangle^0 - P\langle x - x_2 \rangle - \frac{q}{2}\langle x - x_3 \rangle^2 \qquad (6.5)$$

From Equation (6.1), thus:

$$\frac{d^2 y(x)}{dx^2} = -\frac{1}{EI}[R_A x + M\langle x - x_1 \rangle^0 - P\langle x - x_2 \rangle - \frac{q}{2}\langle x - x_3 \rangle^2]$$

$$\frac{dy(x)}{dx} = -\frac{1}{EI}[\frac{R_A}{2} x^2 + M\langle x - x_1 \rangle^1 - \frac{P}{2}\langle x - x_2 \rangle^2 - \frac{q}{6}\langle x - x_3 \rangle^3] + C_1$$

$$y(x) = -\frac{1}{EI}[\frac{R_A}{6} x^3 + \frac{M}{2}\langle x - x_1 \rangle^2 - \frac{P}{6}\langle x - x_2 \rangle^3 - \frac{q}{24}\langle x - x_3 \rangle^4] + C_1 x + C_2$$

The unknown constants C_1 and C_2 are determined using the support conditions.

The general procedure of Macaulay's method is as follows:

- Set up a coordinate system as shown in Figure 6.4.
- Calculate the support reactions.
- Express bending moment in a single expression in terms of the step function.
- Integrate the equation of deflection (Equation (6.1)).
- Solve the two integration constants from support conditions.
- Insert the obtained integration constants back into the solution of deflection.

6.3   Key points review

- Under the action of bending moment, the axis of a beam deflects to a smooth and continuous curve.
- The perpendicular displacement of a beam axis away from its original position is called *deflection*.
- Due to deflection, cross-sections of a beam rotate about its neutral axis. The rotation is called *slope*, which approximately equals the first derivative of deflection with respect to the coordinate in the axial direction.
- Curvature of a deflected beam is proportional to the bending moment acting on its cross-section, while inversely proportional to the flexural rigidity of the beam, EI.
- Bending to an arc of a circle occurs when M/EI is a constant (curvature is a constant).
- Deflection of a beam depends on not only M/EI, but also support conditions.
- Deflection of a beam can be reduced by either using a stiffer section or adding intermediate supports.
- Deflection of a beam subjected to multiple loads is a summation of the deflections of the same beam subjected to each of the loads individually. (Superposition does not apply in inelastic problems.)
- The equation of deflection (Equation (6.1)) is only applicable to deflection due to bending. The deflection due to shear forces are, however, comparatively small in most practical cases.

6.4 Examples

6.4.1 Examples of the integration method

EXAMPLE 6.1

Derive expressions for the slope and deflection of a uniform cantilever of length L, loaded with a uniformly distributed load (UDL). The flexural rigidity of the beam is EI. Calculate also the slope and deflection at the free end.

Figure E6.1

[Solution] *This is a simple problem. Integration can be performed on either Equation (6.1) or (6.3). Starting from Equation (6.1), bending moment along the x-axis must be sought first and two support conditions are needed to complete the solution. If Equation (6.3) is used, four support conditions are needed.*

(a) *Expressing bending moment in terms of x and integrating Equation (6.1)*
   (Two boundary conditions are needed. They can be at $x = 0, y = 0$ and at $x = 0, dy/dx = 0$)

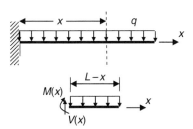

Taking the origin of the coordinate system at the fixed end, the bending moment at an arbitrary section, distance x from the origin, is:

$$M(x) = -q(L-x) \times \frac{L-x}{2} = -\frac{q(L-x)^2}{2}$$

Substituting for $M(x)$ in Equation (6.1):

$$\frac{d^2y}{dx^2} = -\frac{M(x)}{EI} = \frac{q}{2EI}(L^2 - 2Lx + x^2)$$

Since EI is a constant, by direct integration:

$$\frac{dy}{dx} = \frac{q}{2EI} \int (L^2 - 2Lx + x^2)\, dx = \frac{q}{2EI}\left(L^2x - Lx^2 + \frac{x^3}{3}\right) + C_1$$

C_1 is a constant of integration and can be determined using the condition that the slope is zero at the fixed end, that is, at the origin.

Since $\dfrac{dy}{dx} = 0$ at $x = 0$:

$$C_1 = 0$$

Thus:

$$\frac{dy}{dx} = \frac{q}{2EI}\left(L^2 x - Lx^2 + \frac{x^3}{3}\right)$$

Integrating again:

$$y = \frac{q}{2EI}\int\left(L^2 x - Lx^2 + \frac{x^3}{3}\right)dx = \frac{q}{2EI}\left(\frac{L^2}{2}x^2 - \frac{L}{3}x^3 + \frac{1}{12}x^4\right) + C_2$$

C_2 is the second constant of integration, which can be determined using the condition that the deflection is zero at the origin.
Since $y = 0$ at $x = 0$:

$$C_2 = 0$$

Thus, the deflection of the beam is:

$$y = \frac{q}{2EI}\left(\frac{L^2}{2}x^2 - \frac{L}{3}x^3 + \frac{1}{12}x^4\right)$$

(b) *Integration of Equation (6.3)*
(*Four boundary conditions are needed. They can be at $x = 0, y = dy/dx = 0$ and at $x = L, M = V = 0$*):

$$EI\frac{d^4 y}{dx^4} = q$$

Since EI is a constant, by direct integration and using Equation (6.2):

$$EI\frac{d^3 y}{dx^3} = \int q\,dx = qx + C_1 = -V(x)$$

The constant C_1 is determined by using the condition that the shear force is zero at the free end.
Since $V = 0$ at $x = L$:

$$qL + C_1 = -V(L) = 0$$

Thus $C_1 = -qL$.
Then:

$$EI\frac{d^3 y}{dx^3} = qx - qL$$

Integrating the second time and considering Equation (6.1):

$$EI\frac{d^2y}{dx^2} = \frac{q}{2}x^2 - qLx + C_2 = -M(x)$$

Since $M = 0$ at $x = L$:

$$\frac{q}{2}L^2 - qLL + C_2 = -M(L) = 0$$

Thus, $C_2 = \frac{qL^2}{2}$. Then:

$$EI\frac{d^2y}{dx^2} = \frac{q}{2}x^2 - qLx + \frac{qL^2}{2}$$

Integrating the third time:

$$EI\frac{dy}{dx} = \frac{q}{6}x^3 - \frac{qL}{2}x^2 + \frac{qL^2}{2}x + C_3$$

Since $\frac{dy}{dx} = 0$ at $x = 0$:

$$C_3 = 0.$$

Thus, the expression of slope for the beam is:

$$EI\frac{dy}{dx} = \frac{q}{6}x^3 - \frac{qL}{2}x^2 + \frac{qL^2}{2}x$$

or

$$\frac{dy}{dx} = \frac{q}{2EI}\left(\frac{1}{3}x^3 - Lx^2 + L^2x\right)$$

Finally:

$$EIy = \frac{q}{24}x^4 - \frac{qL}{6}x^3 + \frac{qL^2}{4}x^2 + C_4$$

Since $y = 0$ at $x = 0$:

$$C_4 = 0.$$

The deflection of the beam is therefore:

$$y = \frac{1}{EI}\left(\frac{q}{24}x^4 - \frac{qL}{6}x^3 + \frac{qL^2}{4}x^2\right),$$

$$= \frac{q}{2EI}\left(\frac{1}{12}x^4 - \frac{L}{3}x^3 + \frac{L^2}{2}x^2\right)$$

which is the same as that obtained from (a).

At the free end $x = L$, we have

Slope at the free end:

$$\theta_L = \frac{dy}{dx}\bigg|_{x=L} = \frac{q}{2EI}\left(L^2 L - LL^2 + \frac{L^3}{3}\right) = \frac{qL^3}{6EI}$$

Deflection at the free end:

$$y|_{x=L} = \frac{q}{2EI}\left(\frac{L^2}{2}L^2 - \frac{L}{3}L^3 + \frac{1}{12}L^4\right) = \frac{qL^4}{8EI}$$

The slope and deflection are identical to those shown in Table 6.2.

EXAMPLE 6.2
A simply supported beam is loaded with a concentrated force as shown in Figure E6.2. The flexural rigidity EI is constant. Find the expression of deflection of the beam.

Figure E6.2

[Solution] *The concentrated load applied along the beam results in two different expressions of bending moment. The integration of the equation of deflection must proceed for within AC and CB, which generates four unknown constants of integration. The four conditions that can be used to determine the constants are $y(x=0) = 0$, $y(x=L) = 0$, and the continuity of deflection and slope at C.*

Consider the entire beam and take moment about the right-hand-side end, which yields:

$$R_A L - Pb = 0$$

$$R_A = \frac{Pb}{L}$$

For a cross-section within AC, the bending moment is:

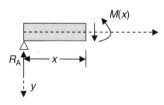

$$M(x) = R_A x = \frac{Pb}{L}x$$

From Equation (6.1):

$$EI\frac{d^2y}{dx^2} = -\frac{Pb}{L}x$$

So:

$$EI\frac{dy}{dx} = -\frac{Pb}{2L}x^2 + C_1 \qquad\qquad\text{(E6.2a)}$$

$$EIy = -\frac{Pb}{6L}x^3 + C_1x + C_2 \qquad\qquad\text{(E6.2b)}$$

Since $y = 0$ at $x = 0$:

$$C_2 = 0$$

Thus at $x = a$:

$$EIy|_{x=a} = -\frac{Pb}{6L}a^3 + C_1a$$

$$EI\frac{dy}{dx}\bigg|_{x=a} = -\frac{Pb}{2L}a^2 + C_1$$

For a cross-section within CB, the bending moment is:

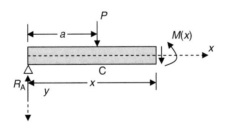

$$M(x) = R_Ax - P(x-a) = \frac{Pb}{L}x - P(x-a)$$

$$= Pa - \frac{Pa}{L}x$$

From Equation (6.1):

$$EI\frac{d^2y}{dx^2} = -Pa + \frac{Pa}{L}x$$

So:

$$EI\frac{dy}{dx} = -Pax + \frac{Pa}{2L}x^2 + C_3$$

$$EIy = -\frac{Pa}{2}x^2 + \frac{Pa}{6L}x^3 + C_3x + C_4$$

Since $y = 0$ at $x = L$:

$$-\frac{PaL^2}{3} + C_3 L + C_4 = 0 \tag{E6.2c}$$

At $x = a$:

$$EI\frac{dy}{dx}\bigg|_{x=a} = -Pa^2 + \frac{Pa}{2L}a^2 + C_3 \tag{E6.2d}$$

$$EIy|_{x=a} = -\frac{Pa}{2}a^2 + \frac{Pa}{6L}a^3 + C_3 a + C_4 \tag{E6.2e}$$

Equating deflections of both AC and CB at $x = a$ yields (Equations (E6.2b) and (E6.2e)):

$$-\frac{Pb}{6L}a^3 + C_1 a = -\frac{Pa}{2}a^2 + \frac{Pa}{6L}a^3 + C_3 a + C_4 \tag{E6.2f}$$

Equating slopes of both AC and CB at $x = a$ yields (Equations (E6.2a) and (E6.2d)):

$$-\frac{Pb}{2L}a^2 + C_1 = -Pa^2 + \frac{Pa}{2L}a^2 + C_3 \tag{E6.2g}$$

The solution of Equations (E6.2c), (E6.2f) and (E6.2g) gives:

$$C_1 = \frac{Pb}{6L}(L^2 - b^2)$$

$$C_3 = \frac{Pa}{6L}(2L^2 + a^2)$$

$$C_4 = \frac{Pa^3}{6}$$

With these constants, the expressions of deflection of the beam are:

$$y = \frac{Pbx}{6EIL}[(L^2 - b^2) - x^2] \qquad\qquad 0 \le x \le a$$

$$y = \frac{Pb}{6EIL}\left[\frac{L}{b}(x-a)^3 + (L^2 - b^2)x - x^3\right] \qquad a \le x \le L$$

It can be concluded from the above example that when applied loads have abrupt changes, including concentrated forces, moments and patched distributed loads, along a beam, using the direct integration method normally involves solving simultaneous linear algebraic equations. To avoid this, Macaulay's method could be a better option.

6.4.2 Examples of the superposition method

The superposition method is quite useful when a complex load can be resolved into a super-position of several simple loads, and the deflections due to these simple loads are known. This method is particularly useful for calculating deflections and slopes at given points along a beam.

EXAMPLE 6.3
A uniform cantilever is subjected to a uniformly distributed pressure and a concentrated force applied at the free end (Figure E6.3). Calculate the deflection and slope of the beam at the free end.

Figure E6.3

[Solution] *This is a simple question showing the basic principle of superposition. The beam can be assumed as under a combined action of the uniformly distributed pressure and the point load applied at the free end. The deflection and slope of the beam under a single action of either the pressure or the point load can be found from Table 6.2.*
The problem can be resolved into two simple problems as:

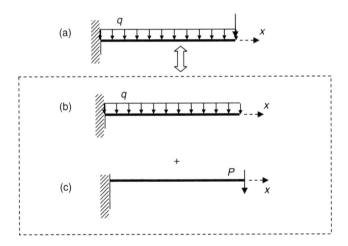

From Table 6.2 the free end deflection and rotation due to the uniform pressure are, respectively:

$$y_{(a)} = \frac{qL^4}{8EI}$$

$$\theta_{(a)} = \frac{qL^3}{6EI}$$

The deflection and slope at the free end due to the concentrated force can also been found from Table 6.2 (let $b = 0$ and $a = L$ in Case 1), which are:

$$y_{(b)} = \frac{PL^3}{3EI}$$

$$\theta_{(b)} = \frac{PL^2}{2EI}$$

Thus, the total deflection and slope of the beam at the free end are, respectively:

$$y_{Total} = y_{(a)} + y_{(b)} = \frac{qL^4}{8EI} + \frac{PL^3}{3EI}$$

$$\theta_{Total} = \theta_{(a)} + \theta_{(b)} = \frac{qL^3}{6EI} + \frac{PL^2}{2EI}$$

EXAMPLE 6.4
Calculate the free end deflection of the beam shown in Figure E6.4. The beam has a uniform flexural rigidity.

Figure E6.4

[Solution] *To use the solution presented in Table 6.2, the UDL is extended to the fixed end (Case (a)) and is effectively removed by applying an upward UDL of equal magnitude between A and B (Case (b)).*

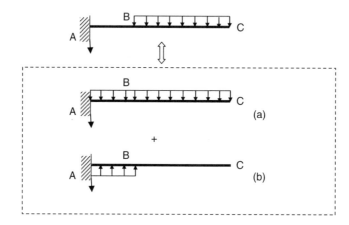

From Table 6.2, the deflection and the slope at the free end of Case (a) are, respectively:

$$y_C^{(a)} = \frac{ql^4}{8EI}$$

$$\theta_C^{(a)} = \frac{ql^3}{6EI}$$

For Case (b) since the segment between B and C is not loaded and the right-hand-side end is completely free, the slope at C is identical to the slope at B. The deflection at the free end equals the deflection at B plus the product of the slope at B and the length of BC:

$$\theta_C^{(b)} = \theta_B^{(b)} = -\frac{qa^3}{6EI}$$

$$y_C^{(b)} = y_B^{(b)} + \theta_B^{(b)}(l-a) = -\frac{qa^4}{8EI} - \frac{qa^3}{6EI}(l-a)$$

The total deflection and slope at the free end are, respectively:

$$y_C = y_C^{(a)} + y_C^{(b)} = \frac{ql^4}{8EI} - \frac{qa^4}{8EI} - \frac{qa^3}{6EI}(l-a) = \frac{q}{24EI}(3l^4 - 4la^3 + a^4)$$

$$\theta_C = \theta_C^{(a)} + \theta_C^{(b)} = \frac{ql^3}{6EI} - \frac{qa^3}{6EI} = \frac{q}{6EI}(l^3 - a^3)$$

EXAMPLE 6.5
A simply supported uniform beam is subjected to a uniform pressure from the mid-span to the right-hand-side support as shown in Figure E6.5. Use the superposition method to calculate the mid-span deflection.

Figure E6.5

[Solution] *There are no solutions in Table 6.2 that can be used directly. However, the UDL can be taken as a sum of an infinite number of small concentrated forces acting on the beam. The deflections at the mid-span due to these concentrated forces can be found from Table 6.2. The summation of the infinite number of deflections can be calculated in the form of integration.*

The beam shown above is subjected to a concentrated force, qda, at a distance a away from the left-hand-side support. From Table 6.2 (Case 3), the following mid-span deflection can be obtained:

$$dy_m = \frac{qda(L-a)L/2}{6LEI}[L^2 - (L-a)^2 - \frac{L^2}{4}]$$

$$= \frac{q}{EI}[\frac{L^2}{16}(L-a) - \frac{1}{12}(L-a)^3]da$$

when P is replaced by qda, I replaced by L, x replaced by $L/2$, and b is replaced by $L-a$. Thus, the total mid-span deflection due to the distributed load is:

$$y_m = \int_{L/2}^{L} \frac{q}{EI}[\frac{L^2}{16}(L-a) - \frac{1}{12}(L-a)^3]da = \frac{5qL^4}{768EI}$$

EXAMPLE 6.6
Find the deflection for the uniformly loaded, two-span continuous beam shown in Figure E6.6. EI is constant.

Figure E6.6

[Solution] *This is a statically indeterminate beam whose reactions cannot be determined using the equilibrium conditions. Superposition method can, sometimes, be used conveniently to calculate the reactions. The beam shown in Figure E6.6 can be taken as the superposition of a beam subjected to the uniform downward pressure and the same beam subjected to an unknown concentrated upward force at the mid-span. The combined action of the two loads results in a zero deflection at the mid-span of the beam.*

From Table 6.2, the mid-span deflections of the beam due to the UDL and the concentrated force are, respectively:

$$y_{UDL} = \frac{5q(2L)^4}{384EI} = \frac{5qL^4}{24EI}$$

$$y_P = -\frac{P(2L)^3}{48EI} = -\frac{PL^3}{6EI}$$

The sum of these two deflections must be zero because the beam is roller-pinned at the mid-span. So:

$$\frac{5qL^4}{24EI} - \frac{PL^3}{6EI} = 0$$

$$P = \frac{5qL}{4}$$

The deflection of the beam is equal to the deflection due to the uniform load plus the deflection of the beam subjected to an upward concentrated force, $5qL/4$, at the mid-span. From Table 6.2, for example, the deflection of the left-hand-side span is:

$$y_{UDL}(x) = \frac{qx}{24EI}(l^3 - 2lx^2 + x^3) = \frac{qx}{24EI}[(2L)^3 - 2(2L)x^2 + x^3]$$

$$= \frac{qx}{24EI}(8L^3 - 4Lx^2 + x^3)$$

$$y_P(x) = \frac{Pbx}{6lEI}(l^2 - b^2 - x^2) = \frac{PLx}{6(2L)EI}[(2L)^2 - L^2 - x^2]$$

$$= \frac{5qx}{48EI}[3L^3 - x^2L]$$

$$y(x) = y_{UDL}(x) - y_P(x) = \frac{q}{48}(2x^4 - 3Lx^3 + L^3x) \qquad 0 \le x \le L$$

Due to symmetry, the deflection of the right-hand-side span is numerically identical to the above though the expression of the curve is different ($y_P(x)$ for $L \le x \le 2L$ should be used in the above superposition).

6.4.3   Examples of Macaulay's method

EXAMPLE 6.7
A uniform beam 16 ft long is simply supported at its ends and carries a uniform distributed load of $q = 0.5$ ton/ft between B and C, which are 3 ft and 11 ft from A, respectively. The beam also carries a concentrated force of $P = 6$ tons 13 ft from A. If $E = 13,400$ tons/in.² and $I = 204.8$ in.⁴, obtain the deflection of the beam and calculate the mid-span deflection.

Figure E6.7

[Solution] *The change of load patterns can be dealt with by the step functions (Equation (6.4)) and the general expression of bending moment (Equation (6.5)). When using Macaulay's method with distributed loads it is essential that the distributed load is continued to the end of the beam. In this example, the distributed load is extended to cover the extra range CD. To remove the additional loading an upward pressure of the same magnitude must also be applied in the same range.*

The equivalent loading of the beam is shown below:

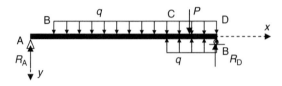

To use Equation (6.5), the reaction force at A must be obtained first.
   Taking moment about D (Figure E6.7)

$$16R_A = 0.5 \times 8 \times 9 + 6 \times 3$$

$$R_A = 3.375 \text{ tons}$$

Hence from Equation (6.5), for an arbitrary section at a distance x from A:

$$M(x) = R_A x - P \langle x - 13 \rangle - \frac{q}{2} \langle x - 3 \rangle^2 + \frac{q}{2} \langle x - 11 \rangle^2$$

$$= 3.375x - 6 \langle x - 13 \rangle - \frac{0.5}{2} \langle x - 3 \rangle^2 + \frac{0.5}{2} \langle x - 11 \rangle^2$$

From Equation (6.1):

$$EI\frac{d^2y}{dx^2} = -M(x)$$

$$= -3.375x + 6 \langle x - 13 \rangle + \frac{0.5}{2} \langle x - 3 \rangle^2 - \frac{0.5}{2} \langle x - 11 \rangle^2$$

Integrating once:

$$EI\frac{dy}{dx} = -\frac{3.375}{2}x^2 + 3 \langle x - 13 \rangle^2 + \frac{1}{12} \langle x - 3 \rangle^3 - \frac{1}{12} \langle x - 11 \rangle^3 + C_1$$

Integrating twice:

$$EIy = -\frac{3.375}{6}x^3 + \langle x - 13 \rangle^3 + \frac{1}{48} \langle x - 3 \rangle^4 - \frac{1}{48} \langle x - 11 \rangle^4 + C_1 x + C_2$$

The two constants of integration can be determined by introducing support conditions.

   At $x = 0$, $y = 0$,

$C_2 = 0$ (use the properties of step function defined in Equation (6.4))

At $x = 16$, $y = 0$.

$$0 = -\frac{3.375}{6} \times 16^3 + (16-13)^3 + \frac{1}{48}(16-3)^4 - \frac{1}{48}(16-11)^4 + 16C_1$$

$C_1 = 105.94$

The deflection of the beam is:

$$y(x) = \frac{1}{EI}[-0.5625x^3 + (x-13)^3 + \frac{1}{48}(x-3)^4 - \frac{1}{48}(x-11)^4 + 105.94x]$$

At the mid-span, $x = 8''$:

$$y_{mid} = \frac{12^2}{13400 \times 204.8}[-0.5625 \times 8^3 + \frac{1}{48}5^4 + 105.94 \times 8] = 0.03\,\text{ft}$$

$$= 0.361\,\text{in}.$$

EXAMPLE 6.8
A built-in beam of length L carries a concentrated load P at distance a from the left-hand-side end (Figure E6.8). Obtain the deflection of the beam and calculate the fixed end moments.

Figure E6.8

[Solution] *This is a statically indeterminate problem. The reaction forces at A and B cannot be solved by static equilibrium. At both ends unknown bending moments and shear forces exist. Before these forces are found, the general expression of bending moment is expressed in terms of these unknown reactions that can be determined subsequently by introducing support conditions.*

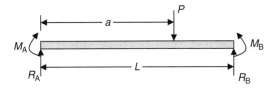

Assume that the two reaction forces at A are, respectively, R_A and M_A. From Equation (6.5), the general expression of bending moment is:

$$M(x) = R_A x + M_A (x-0)^0 - P(x-a) = R_A x + M_A - P(x-a)$$

From Equation (6.1):

$$EI\frac{d^2y}{dx^2} = -M(x) = -R_Ax - M_A + P\langle x-a\rangle$$

Integrating once:

$$EI\frac{dy}{dx} = -\frac{R_A}{2}x^2 - M_Ax + \frac{P}{2}\langle x-a\rangle^2 + C_1 \qquad\text{(E6.8a)}$$

At the fixed end:

$$x = 0, \quad \frac{dy}{dx} = 0,$$

Thus from Equation (E6.8a):

$$C_1 = 0$$

Integrating twice:

$$EIy = -\frac{R_A}{6}x^3 - \frac{M_A}{2}x^2 + \frac{P}{6}\langle x-a\rangle^3 + C_2 \qquad\text{(E6.8b)}$$

Again at the fixed end:

$$x = 0, \quad y = 0$$

From Equation (E6.8b):

$$C_2 = 0$$

The expression of deflection in terms of the unknown reaction forces at A is:

$$y(x) = \frac{1}{EI}[-\frac{R_A}{6}x^3 - \frac{M_A}{2}x^2 + \frac{P}{6}\langle x-a\rangle^3] \qquad\text{(E6.8c)}$$

The reaction force, R_A, and the fixed end moment, M_A, can be determined through the introduction of the support conditions at B.

At $x = L$:

$$\frac{dy}{dx} = 0 \text{ and } y = 0$$

From Equation (E6.8a):

$$0 = -\frac{R_A}{2}L^2 - M_AL + \frac{P}{2}(L-a)^2$$

From Equation (E6.8b):

$$0 = -\frac{R_A}{6}L^3 - \frac{M_A}{2}L^2 + \frac{P}{6}(L-a)^3$$

Solving the above simultaneous equations in terms of R_A and M_A yields:

$$M_A = -\frac{Pa(L-a)^2}{L^2}$$

$$R_A = \frac{P(L-a)^2}{L^3}(2a+L)$$

Once M_A and R_A are found, M_B and R_B can be easily obtained by considering the equilibrium of the entire beam, which are, respectively:

$$M_B = -\frac{P(L-a)a^2}{L^2}$$

$$R_B = \frac{Pa^2}{L^3}(3L-2a)$$

Substituting the above solutions into Equation (E6.8c) yields the deflection of the beam.

6.5   Conceptual questions

1. Describe what is meant by a 'simply supported' support.
2. Describe what is meant by a 'fixed end' support.
3. What is the support condition of a free end?
4. What are deflection, slope and curvature of a beam?
5. How are deflection, slope and curvature related to each other for a beam in bending?
6. What is the difference between axial stiffness and flexural rigidity of a member?
7. What is the difference between torsional stiffness and flexural rigidity of a member?
8. Under what condition will a beam bend in to a circular arc?
9. Is the curvature of a beam zero at the location where bending moment is zero?
10. Does maximum deflection always occur at the position where slope is zero?
11. Does maximum deflection always occur at the position where bending moment is maximum?
12. If the integration method is applied to Equation (6.1) to find deflection of the beam shown in the figure below, how many constants of integration in total are to be determined from imposing support and continuity conditions? And what are the conditions?

Figure Q6.12

13. If two beams have the same length and flexural rigidity and are subjected to the same external loads, are the deflections of the two beams identical, and why?

6.6 Mini test

Problem 6.1:   The cantilever shown in Figure P6.1 has a rectangular cross-section. With all the conditions staying the same, except that the depth of the beam is doubled, complete the following statements:

Figure P6.1

(a) The maximum normal stress is now _____ times the maximum normal stress in the original beams.

(b) The maximum shear stress is now _____ times the maximum shear stress in the original beams.

(c) The maximum deflection is now _____ times the maximum deflection of the original beam.

(d) The maximum slope is now _____ times the maximum slope of the original beam.

Problem 6.2:   Calculate the mid-span deflection of the beams shown in Figure P6.2 using the superposition method and the solutions from Table 6.2.

Figure P6.2

Problem 6.3:   Figure P6.3 shows a horizontal beam freely supported at its ends by the free ends of two cantilevers. If the flexural rigidity, EI, of the cantilevers is twice that of the beam, calculate the deflection at the center of the beam.

Figure P6.3

Problem 6.4: Determine the deflection along the beam, and the magnitude of deflection at D of the beam shown in Figure P6.4. The flexural rigidity (*EI*) is 1.0 MN m².

Figure P6.4

Problem 6.5: Determine the deflection curve of the statically indeterminate beam shown in Figure P6.5. The beam has a constant *EI*.

Figure P6.5

7 Complex stresses

In a practical design, a structure is usually subjected to a combination of different types of loading that generate different types of stresses within the structure. For example, the stress field in a beam–column joint is very complex, with combinations of bending, shearing and contact stresses (Figure 7.1(a)). If a cut (plane) is taken through a point, the stress on the plane is usually different to the stress on a different plane through the same point, not just in terms of magnitude but also direction. On an arbitrary plane through a point, a general (σ) stress can always be resolved into three independent components that are perpendicular to each other (Figure 7.1(b)). The three components include a normal stress (σ_n), which is perpendicular to the plane, two shear stresses (τ_1 and τ_2), which are parallel to the plane and perpendicular to each other. The stresses at a point inside the joint are best presented by the stresses acting on an infinitesimal cubic element taken around the point. The element has six faces (planes) that are

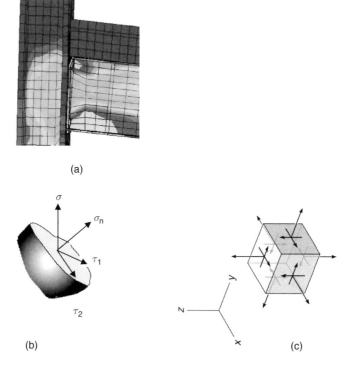

(a)

(b)

(c)

Figure 7.1

either perpendicular or parallel to each other. On each of the faces there are three independent stresses, including two shear stresses and a normal stress.

Figure 7.2 shows all the stresses at a point in a material, which is sufficient and necessary to represent the *state of stress* at the point.

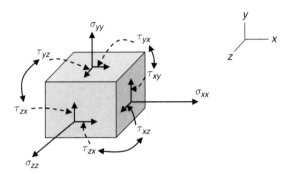

Figure 7.2

- When a structure is subjected to external loads, the state of stress is, in general, different at different points within the structure.
- At a point in the structure, the stress in one direction is usually different to the stress in a different direction.
- State of stress (Figure 7.1(c)) shows stresses acting on six different planes at a point. Therefore, when we say that we know a stress, it means that we know not only the magnitude and direction of the stress, but also the plane on which the stress acts.
- Since the cubic element has infinitesimal dimensions in the three co-ordinate directions, the normal stresses acting on any two faces that are parallel to each other are equal but in opposite directions. On any two planes that are perpendicular to each other, the shear stresses perpendicular to the intersection of the two planes are equal, but in an opposite sense, i.e., are either towards or away from the intersection line. At a point, therefore, there are only six independent stresses, i.e., σ_{xx}, σ_{yy}, σ_{zz}, $\tau_{xy} (= \tau_{yx})$, $\tau_{xz} (= \tau_{zx})$ and $\tau_{yz} (= \tau_{zy})$ (Figure 7.2).
- At any point within a material, if the two shear stresses are zero on a plane, this plane is called *principal plane*. The normal stress acting on the principal plane is called *principal stress*, and its direction is called *principal direction*. If a cubic element is chosen such that all the faces of the cube are free of shear stresses, the element is called *principal element*.

7.1 Two-dimensional state of stress

In some cases, the stresses relative to a particular direction are sufficiently small compared to the stresses relative to the other two directions. Typical example problems include stresses in a thin plate subjected to in-plane loadings (Figure 7.3(a)) and in a thin-walled vessel under internal pressure (Figure 7.3(b)) or torsion (Figure 7.3(c)). Suppose that the small stress is related

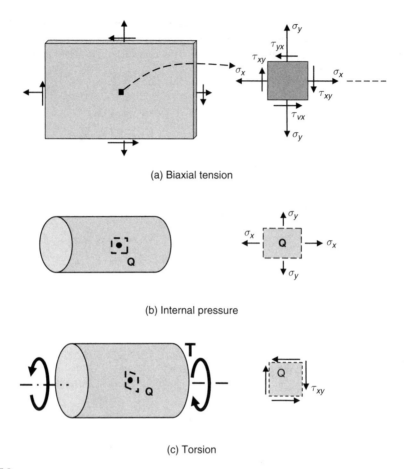

(a) Biaxial tension

(b) Internal pressure

(c) Torsion

Figure 7.3

to the z direction and is ignored, the three-dimensional state of stress can be reduced to a two-dimensional one. Since the remaining stresses lie in a plane, the simplified two-dimensional problems are called *plane problems*. For the thin plate subjected to in-plane loads, Figure 7.3(a) shows the two-dimensional state of stress. For a thin-walled cylinder subjected to internal pressure or torsion, the states of stress are shown by Figures 7.3(b) and 7.3(c).

In Figure 7.3, the normal stresses (σ_x and σ_y have a single subscript index that indicates the coordinate axis the stresses are parallel to. The first subscript index of a shear stress (τ_{xy} or τ_{yx}) denotes the direction of the normal of the plane on which the stress acts, while the second index denotes the axis to which the shear stress is parallel. Since the two-dimensional element is infinitesimal, τ_{xy} is numerically equal to τ_{yx}.

7.1.1 Sign convention of stresses

The stresses shown in Figure 7.3 are all defined as positive in the chosen coordinate system, where the following sign conventions are followed:

- Tensile and compressive stresses are always defined, respectively, as positive and negative.

- Positive shear stresses are defined the same as the positive shear forces defined in Section 4.4.

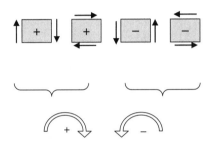

7.1.2 Analytical method

Once the stress components that align with a typical x–y coordinate system are found (Figure 7.2), transforming the stresses into another coordinate system is sometimes necessary. Two key reasons that we may want to calculate stresses in a different coordinate system include the following:

- To determine the stress in an important direction, for example, stresses normal and parallel to the plane of a weld (Figure 7.4(b));
- To determine the maximum normal stress or maximum shear stress at a point. These stresses may not necessarily align with the chosen coordinate directions.

(a) *Stresses on an arbitrarily inclined plane*

To further understand the stresses on an arbitrarily inclined plane (cut) through a point in a material, consider the two elements taken around the point in Figure 7.4.

Since the elements are taken at the same point, we might take them for the same state of stress, but measured in a different coordinate system. The new coordinate system, x'–y', is defined by a rotation θ of both the coordinate axes from their original directions (Figure 7.5). The relationship between the states of stress in terms of the original system and the rotated system can be best presented by considering the equilibrium of the wedge of material (Figure 7.6) taken from Figure 7.5(b). The normal stress, σ_θ, and shear stress, τ_θ, acting on the inclined plane are, respectively, the normal stress, $\sigma_{x'}$, and shear stress, $\tau_{x'y'}$, in the rotated coordinate system, while the stresses acting on the vertical and horizontal sides of the wedge are identical to those acting on the vertical and horizontal sides of the element (Figure 7.5(a)) in the original coordinate system.

Figure 7.4

Figure 7.5

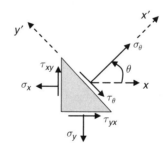

Figure 7.6

The equilibrium of the wedge yields:

$$\sigma_\theta = \sigma_x \cos^2\theta + \sigma_y \sin^2\theta - \tau_{xy}\sin 2\theta$$

$$= \frac{\sigma_x + \sigma_y}{2} + \frac{\sigma_x - \sigma_y}{2}\cos 2\theta - \tau_{xy}\sin 2\theta \qquad (7.1)$$

$$\tau_\theta = \left(\frac{\sigma_x - \sigma_y}{2}\right)\sin 2\theta + \tau_{xy}\cos 2\theta$$

where an anticlockwise angle from the x-axis is defined as positive.

(b) *Principal stresses*

From Equation (7.1), the stresses on an inclined plane change as the value of θ changes. It means that on different planes taken by cutting through the point the stresses are generally different. It is natural to think that there are special planes on which the normal stress reaches either maximum or minimum (maximum compressive stress) algebraically. The maximum and minimum normal stresses are both called *principal stresses*. When a normal stress is either maximum or minimum, the plane on which the stress acts is always free of shear stress. In a two-dimensional stress system, there are two principal stresses, that is, the maximum and the minimum normal stresses at a point, as shown in Figure 7.7. The principal stresses can be calculated as follows:

$$\sigma_{max} = \sigma_1$$

$$= \frac{1}{2}[(\sigma_x + \sigma_y) + \sqrt{(\sigma_x - \sigma_y)^2 + 4\tau_{xy}^2}]$$

$$(7.2)$$

$$\sigma_{min} = \sigma_2$$

$$= \frac{1}{2}[(\sigma_x + \sigma_y) - \sqrt{(\sigma_x - \sigma_y)^2 + 4\tau_{xy}^2}]$$

(c) *The directions of principal stresses*

The angle between a principal stress and the x-axis can be calculated as follows:

$$\tan 2\theta = \frac{-2\tau_{xy}}{\sigma_x - \sigma_y}$$

$$(7.3)$$

$$\theta = \frac{1}{2}\tan^{-1}\left(\frac{-2\tau_{xy}}{\sigma_x - \sigma_y}\right)$$

Since the two principal stresses are perpendicular to each other, the direction of the second principal stress is $\theta + 90°$.

Figure 7.7

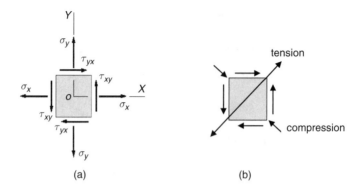

Figure 7.8

In a plane problem, there are two principal stresses (Equation (7.2)) and two associated directions. The easiest way to relate the stresses to their respective directions is based on the following simple observation.

In Figure 7.8, the shear stresses τ_{xy} generates tension in one diagonal direction and compression in the other, which suggests that combined with actions of σ_x and σ_y, the normal stress in the direction of the tension diagonal is *more tensile* or larger than that in the direction of the compression diagonal. Hence, it can be concluded that the direction of σ_1 is related to where the shear stresses are pointing to.

At a point in a material, a normal stress is a principal stress if:

- the stress is either the maximum tensile stress or the maximum compressive stress at the point;

or

- the plane on which the normal stress acts is free of shear stresses. The plane is one of the principal planes.

(d) *Maximum shear stress*

Following the same argument as for the existence of maximum normal stresses, there exist special planes on which shear stress reaches maximum or minimum. (They have equal magnitudes in an opposite sense.) Figure 7.9 shows a concrete cylinder under compression. The cylinder fails due to maximum shearing at about 45° to the axial of compression. The cylinder may fail along the other diagonal direction under the same compression due to an equal shear stress of opposite sense. By observation, the plane perpendicular to the axial direction is a principal plane since there is no shear stress acting on this plane. The maximum shear stress acts on the plane that is 45° away from the principal plane. From Equation (7.1), the maximum shear stress can be obtained by replacing σ_x and σ_y with σ_1 and σ_2, respectively:

$$\tau_{max} = \frac{\sigma_1 - \sigma_2}{2} \tag{7.4a}$$

and from Equation (7.2):

$$\tau_{max} = \frac{1}{2}\sqrt{(\sigma_x - \sigma_y)^2 + 4\tau_{xy}^2} \tag{7.4b}$$

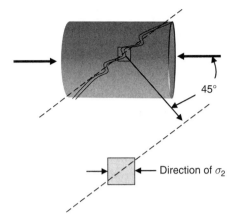

Figure 7.9

7.1.3 Graphic method

Mohr's circle illustrates principal stresses and stress transformations via a graphical format, that is, a graphic representation of Equations (7.1–7.4). The circle is plotted in a plane coordinate system where the horizontal axis denotes normal stress. The vertical coordinate denotes the shear stress on the same plane (Figure 7.10). While plotting a Mohr's circle a sign convention, for example, the one defined in Section 7.1.1, must be followed. Here we take tensile stresses and shear stresses that would turn an element clockwise as positive. Figure 7.10 shows how the stresses acting on an element are related to a Mohr's circle.

The two principal stresses are shown by the σ coordinates of the two intersections of the circle with the horizontal axis (where shear stresses are zero). The vertical coordinates of either the highest or the lowest point on the circle denote the maximum magnitude of shear stress that is also equal to the radius of the circle. If the state of stress at a point is known, that is, σ_x, σ_y and τ_{xy} are known, the following steps can be followed to plot a Mohr's circle:

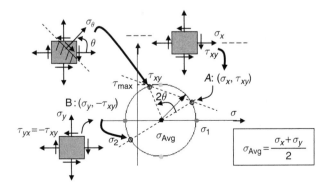

Figure 7.10

- Set up a co-ordinate system where the horizontal axis is the normal stress axis and the vertical axis is the shear stress axis; positive directions of the axes take upwards and to the right.
- Locate two points, A and B, related to, respectively, the stresses on the right and upper faces of a state-of-stress element, with respective coordinates (σ_x, τ_{xy}) and $(\sigma_y, -\tau_{xy})$, in the $\sigma-\tau$ co-ordinate system and connect the two points by a straight line. The sign convention defined in Section 7.1.1 must be followed to locate the two points. The intersection of the straight line with the σ-axis is marked 'O' and is at a distance of $\sigma_{ave} = (\sigma_x + \sigma_y)/2$ away from the origin.
- With its centre at 'O', draw a circle passing through points A and B.
- Measure σ coordinates of the two intersections of the circle with the σ-axis to obtain the two principal stresses.
- Measure the radius of the circle to obtain the maximum shear stress.
- To determine the magnitudes of the stresses acting on an inclined plane, $\theta°$ away from the right-hand-side face, measure an angle of $2\theta°$ from OA and take the coordinates of the intersection with the circle.
- The horizontal and vertical coordinates of the intersection are, respectively, the normal and shear stresses on the inclined plane.

7.2   Key points review

7.2.1   Complex stress system

- At a point in a material, there are six independent stress components, including three normal stresses and three shear stresses.
- In a two-dimensional case, there are three independent stresses, two normal stresses and one shear stress, at a point of the material.
- A stress usually varies from point to point.
- A stress is uniquely defined by the following three properties:

  - magnitude
  - direction
  - plane (cross-section/cut) on which the stress acts.

- Without knowing any of the three, the stress is not completely defined.
- Principal stresses are normal stresses and include both maximum and minimum compressive stresses.
- In a three-dimensional stress system, there are three principal stresses; while in a two-dimensional system, there are two principal stresses.
- Principal stresses are always perpendicular to each other.
- The plane on which a principal stress acts is free of shear stresses.
- Maximum shear stress is equal to half of the maximum difference between principal stresses.
- Maximum shear stress is always 45° away from a principal stress.

7.2.2 Mohr's circle

- The largest and smallest horizontal coordinates of the circle are, respectively, the two principal stresses σ_1 and σ_2.
- The maximum shear stress is numerically equal to the length of the radius of the circle and also equal to $(\sigma_1 - \sigma_2)/2$.
- An angle difference θ between two planes through a point is represented by a difference of 2θ between the two locations relative to the stresses on the two planes along the Mohr's circle.
- A normal stress equal to $(\sigma_x + \sigma_y)/2$ acts on the planes of maximum shear stresses.
- If $\sigma_1 = \sigma_2$, Mohr's circle degenerates into a point, for which no shear stresses develop at the point.

7.3 Examples

EXAMPLE 7.1

At a point in a masonry structure, the stress system caused by the applied loadings is shown in Figure E7.1(a). Calculate the magnitudes and orientations of the principal stresses at the point. If the stone, of which the structure is made, is stratified and is weak in shear along the planes parallel with the A–A and the allowable shear stress of these planes is 2.3 MPa, is this stress system permissible?

Figure E7.1(a)

[Solution] *This is a question asking for calculation of principal stresses and stresses on an inclined plane. The state of stress at the point is given. The principal stresses and the shear stress on the stratified plane can be obtained by application of Equations (7.1–7.3) or by use of Mohr's circle method.*

(a) Analytical solutions:
Following the sign conventions
$\sigma_x = -10\,\text{MPa}$, $\sigma_y = -2\,\text{MPa}$, $\tau_{xy} = -3\,\text{MPa}$ (sign convention in Section 7.1.1)

From Equations (7.2) and (7.3):

$$\sigma_1 = \frac{1}{2}[(\sigma_x + \sigma_y) + \sqrt{(\sigma_x - \sigma_y)^2 + 4\tau_{xy}^2}]$$

$$= \frac{1}{2}[(-10-2) + \sqrt{(-10+2)^2 + 4 \times 3^2}] = -1.0\,\text{MPa}$$

$$\sigma_2 = \frac{1}{2}[(\sigma_x + \sigma_y) - \sqrt{(\sigma_x - \sigma_y)^2 + 4\tau_{xy}^2}]$$

$$= \frac{1}{2}[(-10-2) - \sqrt{(-10+2)^2 + 4 \times 3^2}] = -11.0\,\text{MPa}$$

$$\tan 2\theta = \frac{-2\tau}{\sigma_x - \sigma_y} = \frac{-2 \times (-3)}{-10 - (-2)} = -\frac{3}{4}$$

$\theta = -18.43°$ (clockwise from the x-axis)

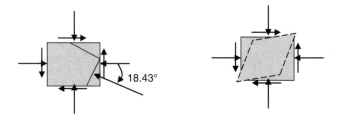

18.43°

Figure E7.1(b)

The above calculations suggest that one of the principal stresses acts in the direction $\theta = -18.43°$ away from the x-axis. Judging by the distortion of the element due to the applied shear stresses (Figure E7.1(b)) and by common sense, it is obvious that the principal stress in the direction of $\theta = -18.43°$ is more compressive than the other principal stress is. Hence, the principal stress acting along $\theta = -18.43°$ is the minimum principal stress, $\sigma_2 = -11$ MPa. The maximum principal stress, $\sigma_1 = -1.0$ MPa, is in the direction perpendicular to σ_2. The principal directions at the point are shown below:

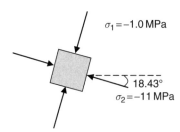

$\sigma_1 = -1.0\,\text{MPa}$

18.43°

$\sigma_2 = -11\,\text{MPa}$

Figure E7.1(c)

The shear stress along plane A–A is calculated from Equation (7.1):

$$\tau_\theta = \left(\frac{\sigma_x - \sigma_y}{2}\right) \sin 2\theta + \tau_{xy} \cos 2\theta$$

$$\tau_{-30°} = -\left(\frac{\sigma_x - \sigma_y}{2}\right) \sin 60° + \tau_{xy} \cos 60°$$

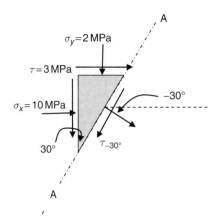

$$= -(\frac{-10+2}{2})\frac{\sqrt{3}}{2} + (-3) \times \frac{1}{2} = 1.96\,\text{MPa}$$

$$\tau_{-30°} < 2.3\,\text{MPa}$$

Hence, the stress system is permissible.

(b) Mohr's circle solution
According to the sign convention we have followed, the stress system shown in Figure E7.1(a) has $(\sigma_x, \tau_{xy}) = (-10\,\text{MPa}, -3\,\text{MPa})$ on the right-hand-side vertical plane and $(\sigma_x, -\tau_{xy}) = (-2\,\text{MPa}, 3\,\text{MPa})$ on the upper horizontal plane, respectively. Thus, the Mohr's circle can be plotted by following the steps described in Section 7.1.3:

- Locate the two points with coordinates $A = (-10, -3)$ and $B = (-2, 3)$, respectively, in the $\sigma-\tau$ plane. Point A represents the right-hand side of the element and Point B represents the top side of the element.
- Connect the two points by a straight line to establish the diameter of the circle. The intersection with the σ-axis is the average of the two normal stresses and marked 'O'.

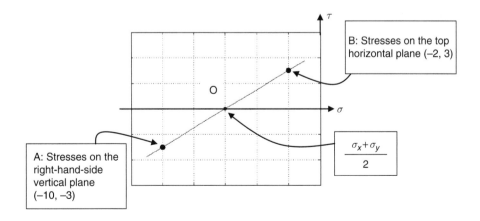

- With its centre at 'O', draw a circle passing through the two points.
- Measure σ coordinates of the two intersections of the circle with the σ-axis, where shear stresses are zero, to obtain the two principal stresses σ_1 and σ_2.

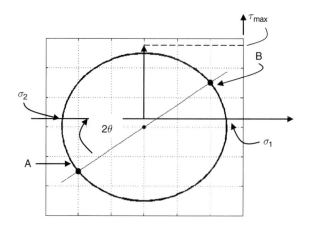

Each division in the figure is equivalent to 2 MPa. Therefore:

$\sigma_1 = 0.5$ division $= -1$ MPa

$\sigma_2 = 5.5$ division $= -11$ MPa

- The angle between the right-hand-side and the upper-side planes of the element (planes A and B) is 90°, while the associated points in the circle is 180° away. This suggests that for two planes having an angle of θ, the points relative to these planes on the Mohr's circle are 2θ away from each other.
  The clockwise angle, 2θ, between plane A and the second principal plane is:

$$\tan 2\theta = \frac{1.5 \text{ division}}{2 \text{ division}} = \frac{3}{4} = 0.75$$

$$\theta = \frac{1}{2} \arctan (0.75) = 18.43°$$

The angle between the normal stress on plane A (the vertical plane on the right-hand side of the element) and the second principal stress is, therefore, 18.43° clockwise.

- Measure the radius of the circle to obtain the maximum shear stress:

$$\tau_{max} = 2.5 \text{ division} = 5 \text{ MPa}$$

or take

$$\tau_{max} = \frac{(\sigma_1 - \sigma_2)}{2} = \frac{-1 + 11}{2} = 5 \text{ MPa}$$

The points associated with the maximum or minimum shear stress are 90° away from the points associated with the two principal stresses on the circle. The maximum shear stress is therefore 45° away from a principal plane.

- Shear stress on the plane 30° clockwise away from the right-hand-side vertical planes.

Point A represents the vertical plane. The shear stress on the plane 30° away from it can be measured from the above figure, which is equal to 1.96 MPa.

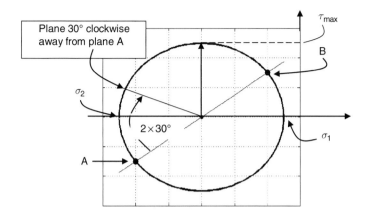

EXAMPLE 7.2

Consider the bar shown in Figure E7.2 under torsional loading. Determine the principal stresses and their orientation.

Figure E7.2

[Solution] *The state of stress of the shaft is determined by the shear stress caused by the torque. On the element taken, there are no normal stresses.*
   Thus:

$$\sigma_x = \sigma_y = 0, \qquad \tau_{xy} = \tau$$

Calculating the magnitudes of σ_1 and σ_2 (Equation (7.2)):

$$\sigma_1 = \frac{\sigma_x + \sigma_y}{2} + \sqrt{\left(\frac{\sigma_x - \sigma_y}{2}\right)^2 + \tau_{xy}^2} = \tau$$

$$\sigma_2 = \frac{\sigma_x + \sigma_y}{2} - \sqrt{\left(\frac{\sigma_x - \sigma_y}{2}\right)^2 + \tau_{xy}^2} = -\tau$$

Calculate the orientation of these principal stresses (Equation (7.3)):

$$\tan 2\theta = \frac{-2\tau_{xy}}{\sigma_x - \sigma_y} = \frac{-\tau}{0} = -\alpha$$

$$2\theta = -90°$$

$$\theta_1 = -45°$$

$$\theta_2 = -45° + 90° = 45°$$

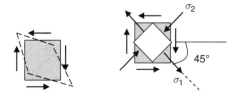

The direction of σ_1 is 45° away from the x-axis clockwise, which can be judged by the above-sketched deformation, where a tensile strain is observed in this direction.

EXAMPLE 7.3
A thin-walled cylinder has an internal diameter of 60 mm and a wall thickness of 1.5 mm. Determine the principal stresses at a point on the outside surface of the generator when the cylinder is subjected to an internal pressure of 6 MPa and a torque, about its longitudinal axis, of 1.0 kN m. If the cylinder is made from plates that are welded along the 45° seams, calculate the normal and shear stresses along the seams.

Figure E7.3

[Solution] *The state of stress is defined by the axial and hoop stresses caused by the internal pressure and the shear stress due to the applied torque. The pressure p acting on the end plate area A is equivalent to an axial force of:*

$$P_x = p \times A(rea) = p \times (\frac{\pi d^2}{4})$$

and the hoop force, P_y, is equal to:

$$P_y = pd/2$$

For the thin-walled cylinder, the cross-sectional area can be calculated approximately by πdt. For the infinitesimal element taken at A (Figure E7.3), the hoop stress is:

$$\sigma_y = \frac{P_y}{t} = \frac{pd}{2t} = \frac{6\,\text{MPa} \times 60 \times 10^{-3}\,\text{m}}{2 \times 1.5 \times 10^{-3}\,\text{m}}$$

$$= 120\,\text{MPa}$$

The axial stress is:

$$\sigma_x = \frac{P_x}{\pi dt} = \frac{pd}{4t} = \frac{6\,\text{MPa} \times 60 \times 10^{-3}\,\text{m}}{4 \times 1.5 \times 10^{-3}\,\text{m}}$$

$$= 60\,\text{MPa}$$

Since the wall of the cylinder is thin, that is, $t << d$, the torsional constant is calculated approximately as:

$$J = \frac{\pi}{32}(D^4 - d^4) = \frac{\pi}{32}d^4(\frac{D^4}{d^4} - 1) = \frac{\pi}{32}d^4[(1 + \frac{2t}{d})^4 - 1] \approx \frac{\pi t}{4}d^3$$

where D is the outside diameter of the cylinder. Considering $D \approx d$, the shear stress due to the torque is then:

$$\tau_{xy} = \frac{T(d/2)}{J} = \frac{1000\,\text{N m} \times 60 \times 10^{-3}\,\text{m}/2}{\pi(1.5 \times 10^{-3}\,\text{m})(60 \times 10^{-3}\,\text{m})^3/4}$$

$$= 118\,\text{MPa}$$

From Equation (7.2):

$$\sigma_1 = \frac{1}{2}[(\sigma_x + \sigma_y) + \sqrt{(\sigma_x - \sigma_y)^2 + 4\tau_{xy}^2}]$$

$$= \frac{1}{2}[(60 + 1200 + \sqrt{(60 - 120)^2 + 4 \times 118^2}] = 212\,\text{MPa}$$

$$\sigma_2 = \frac{1}{2}[(\sigma_x + \sigma_y) - \sqrt{(\sigma_x - \sigma_y)^2 + 4\tau_{xy}^2}]$$

$$= \frac{1}{2}[(60 + 1200) - \sqrt{(60 - 120)^2 + 4 \times 118^2}] = -32\,\text{MPa}$$

The maximum shear stress is (Equation (7.4a)):

$$\tau_{max} = \frac{\sigma_1 - \sigma_2}{2} = \frac{212 - (-32)}{2} = 122\,\text{MPa}$$

Along the 45° seams (Equation (7.1)):

$$\sigma_\theta = \sigma_x \cos^2\theta + \sigma_y \sin^2\theta - \tau_{xy}\sin 2\theta$$

$$= 60\,\text{MPa} \times \cos^2 45° + 120\,\text{MPa} \times \sin^2 45° - 118\,\text{MPa} \times \sin 90°$$

$$= -28\,\text{MPa}$$

$$\tau_\theta = \left(\frac{\sigma_x - \sigma_y}{2}\right)\sin 2\theta + \tau_{xy}\cos 2\theta$$

$$= \left(\frac{60\,\text{MPa} - 120\,\text{MPa}}{2}\right)\sin 90° + 118\,\text{MPa} \times \cos 90°$$

$$= -30\,\text{MPa}$$

EXAMPLE 7.4
The cross-section of a beam, as shown in Figure E7.4, is subjected to a bending moment of $M = 10\,\text{kN m}$ and a shear force of $V = 120\,\text{kN}$. Calculate the principal stresses at points 1, 2, 3 and 4 respectively.

Figure E7.4

[Solution] *The states of stress at the points are determined by the normal stress due to the applied bending moment and the shear stress due to the applied shear force. At 1 and 4, the shear stress is zero, while at 2 the normal stress is zero. At 3 there is a combined action of normal and shear stresses. The states of stress of the four points are, respectively, as follows:*

Calculating the principal stresses of the 4 points is, therefore, a direct application of Equation (7.2) to the above four states of stress.
    The second moment of area of the cross-section is:

$$I = \frac{1}{12}bh^3 = \frac{1}{12} \times 60 \times 10^{-3}\,\text{m} \times (100 \times 10^{-3}\,\text{m})^3 = 500 \times 10^{-8}\,\text{m}^4$$

At location 1, the normal stress due to bending (Equation (5.1a)):

$$\sigma = \frac{My}{I} = \frac{10 \times 10^3\,\text{N m} \times 50 \times 10^{-3}\,\text{m}}{500 \times 10^{-8}\,\text{m}^4} = 100 \times 10^6\,N/m^2 \text{ (Compression)}$$

At 1, since there is no shear stress, the above normal stress is one of the principal stresses. Thus:

$$\sigma_1 = 0$$
$$\sigma_2 = -100\,\text{MPa}$$

At location 2, there is no normal stress (neutral axis). The shear stress is calculated by (Example 5.5):

$$\tau = \frac{3V}{2A} = \frac{3 \times 120 \times 10^3\,\text{N}}{2 \times 60 \times 100 \times 10^{-6}\,\text{m}^2} = 30 \times 10^6\,N/m^2 = 30\,\text{MPa}$$

From Equation (7.2):

$$\sigma_1 = \frac{\sigma_x + \sigma_y}{2} + \sqrt{\left(\frac{\sigma_x - \sigma_y}{2}\right)^2 + \tau_{xy}^2} = 0 + \sqrt{0 + \tau^2} = \tau = 30\,\text{MPa}$$

$$\sigma_2 = \frac{\sigma_x + \sigma_y}{2} - \sqrt{\left(\frac{\sigma_x - \sigma_y}{2}\right)^2 + \tau_{xy}^2} = 0 - \sqrt{0 + \tau^2} = -\tau = -30\,\text{MPa}$$

At location 3 both normal and shear stresses exist. The normal stress due to bending is:

$$\sigma = \frac{My}{I} = \frac{10 \times 10^3\,\text{N m} \times 25 \times 10^{-3}\,\text{m}}{500 \times 10^{-8}\,m^4} = 50 \times 10^6\,N/m^2\ (\text{tension})$$

and the shear stress due to the applied shear force is (Equation 5.5):

Neutral axis

3

25 mm

$$\tau = \frac{VS^*}{bl} = \frac{120 \times 10^3\,\text{N} \times 60 \times 25 \times 37.5 \times 10^{-9}\,\text{m}^3}{60 \times 10^{-3}\,\text{m} \times 500 \times 10^{-8}\,m^4}$$
$$= 22.5 \times 10^6\,N/m^2$$

From Equation (7.2):

$$\sigma_1 = \frac{\sigma_x + \sigma_y}{2} + \sqrt{\left(\frac{\sigma_x - \sigma_y}{2}\right)^2 + \tau_{xy}^2}$$

$$= \frac{50 \times 10^6\,Pa + 0}{2} + \sqrt{(\frac{50 \times 10^6\,Pa - 0}{2})^2 + \tau^2} = 58.6\,\text{MPa}$$

$$\sigma_2 = \frac{\sigma_x + \sigma_y}{2} - \sqrt{\left(\frac{\sigma_x - \sigma_y}{2}\right)^2 + \tau_{xy}^2}$$

$$= \frac{50 \times 10^6\,Pa + 0}{2} - \sqrt{(\frac{50 \times 10^6\,Pa - 0}{2})^2 + \tau^2} = -8.6\,\text{MPa}$$

At location 4, there is no shear stress and the normal stress due to bending at the point is one of the principal stresses:

$$\sigma = \frac{My}{I} = \frac{10 \times 10^3 \, \text{N m} \times 50 \times 10^{-3} \, \text{m}}{500 \times 10^{-8} \, \text{m}^4} = 100 \times 10^6 \, \text{N/m}^2 \text{ (tension)}$$

Thus:

$$\sigma_1 = 100 \, \text{MPa}$$

$$\sigma_2 = 0$$

7.4   Conceptual questions

1. What is meant by 'state of stress' and why is it important in stress analysis?
2. The state of stress at a point of a material is completely determined when

   (a) the stresses on three different planes are specified;
   (b) the stresses on two different planes are specified;
   (c) the stresses on an arbitrary plane are specified;
   (d) none of the above statement is correct.

3. Can both the square elements shown in Figure Q7.3 be used to represent the state of stress at the point of the beam? If yes, which one do you prefer to use and why?

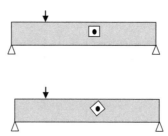

Figure Q7.3

4. Consider a shaft with a constant circular section. If it is subjected to torsion only, at any point within the shaft, normal stress is always zero. (Y/N).
5. What is meant by 'principal stresses', and what is the importance of them?
6. How many principal stresses are there at a point in a two-dimensional stress system?
7. What is meant by 'principal planes', and what is the value of shear stress on these planes?
8. If the principal stresses at a point are known, how can the maximum shear stress at the point be calculated?
9. On the maximum shear stress plane, what values do the normal stresses take?
10. What is the angle between a principal plane and a maximum shear stress plane? Use Mohr's circle to illustrate this.
11. From a Mohr's circle, why do the intersections with the horizontal axis represent the principal stresses at a point?
12. From a Mohr's circle, why does the radius of the circle represent the maximum shear stress at a point?

13. Use Mohr's circle to illustrate the equality of shear stresses on two planes which are perpendicular to each other.
14. The state of stress at a point is shown in Figure Q.7.14. The angle between the *x*-axis and the maximum normal stress is likely to be in the direction of:

Figure Q7.14

    (a) 13.5°
    (b) −96.5°
    (c) 76.5°
    (d) −13.5°.

    (Positive angle is defined as anticlockwise from the *x*-axis)
15. In Equation (7.1), what conclusion can you draw from the sum of σ_1 and σ_2?

7.5 Mini test

Problem 7.1: Among the three states of stress shown in the Figure P7.1, which two are equivalent?

    (a)   (b)   (c)

Figure P7.1

Problem 7.2: A thin-walled hollow sphere of radius *R* with uniform thickness *t* is subjected to an internal pressure *p*. Determine and discuss the state of stress of a point on the outside surface. Is there any shear stress acting at the point? Explain why.

Figure P7.2

Problem 7.3:   Match the states of stress shown in Figure P7.3(b) with the points shown on the two beams in Figure P7.3(a):

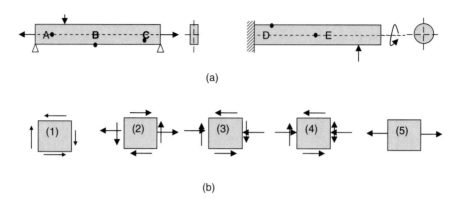

(a)

(1)   (2)   (3)   (4)   (5)

(b)

Figure P7.3

Problem 7.4:   At point *P* on the free edge of a plate, oriented as shown in the Figure P7.4, the maximum shear stress at the point is $4000 \, \text{kN/m}^2$. Use Mohr's circle to find the principal stresses and σ_x at the point in the *x–y* system shown in the figure:

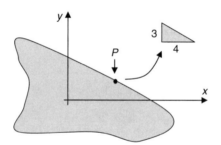

Figure P7.4

Problem 7.5:   A simply supported beam of rectangular cross-section is loaded as shown in Figure P7.5. Determine the states of stress of points 1–5 that are equally spaced across the depth of the beam. Calculate also the principal stresses at these points.

Figure P7.5

8 Complex strains and strain gauges

From Chapter 7, at a point within a three-dimensional material, there are usually six independent stresses as shown in Figure 8.1, that is:

- Three normal stresses: σ_x, σ_y and σ_z
- Three shear stresses: τ_{xy}, τ_{xz} and τ_{yz}

At the same point, due to the action of these stresses, there exist six independent strains, that is:

- Three normal strains: ε_x, ε_y and ε_z
- Three shear strains: γ_{xy}, γ_{xz} and γ_{yz}

For linearly elastic and isotropic materials, the six stresses and the six strains satisfy Hooke's law as follows:

$$\varepsilon_x = \frac{\sigma_x}{E} - v\frac{\sigma_y}{E} - v\frac{\sigma_z}{E}$$
$$= \frac{1}{E}\left[\sigma_x - v(\sigma_y + \sigma_z)\right]$$

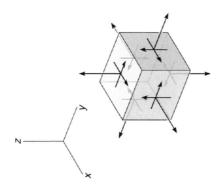

Figure 8.1

$$\varepsilon_y = \frac{\sigma_y}{E} - v\frac{\sigma_x}{E} - v\frac{\sigma_z}{E}$$

$$= \frac{1}{E}[\sigma_y - v(\sigma_x + \sigma_z)]$$

$$\varepsilon_z = \frac{\sigma_z}{E} - v\frac{\sigma_x}{E} - v\frac{\sigma_y}{E}$$

$$= \frac{1}{E}[\sigma_z - v(\sigma_x + \sigma_y)]$$

(8.1a)

$$\gamma_{xy} = \frac{\tau_{xy}}{G}$$

$$\gamma_{xz} = \frac{\tau_{xz}}{G}$$

$$\gamma_{yz} = \frac{\tau_{yz}}{G}$$

(8.1b)

where E, G and v are, respectively, Young's modulus, shear modulus and Poisson's ratio of the material.

In a two-dimensional stress system, where one of the normal stresses, for example, the normal stress in the z direction, is zero, or in most cases negligible, and the shear stresses acting on this particular plane are also zero, the generalized Hooke's law of Equation (8.1) is reduced to the following:

• Strains in terms of stresses:

$$\varepsilon_x = \frac{\sigma_x}{E} - v\frac{\sigma_y}{E} = \frac{1}{E}[\sigma_x - v\sigma_y]$$

$$\varepsilon_y = \frac{\sigma_y}{E} - v\frac{\sigma_x}{E} = \frac{1}{E}[\sigma_y - v\sigma_x]$$

(8.2a)

$$\gamma_{xy} = \frac{\tau_{xy}}{G}$$

• Stresses in terms of strains:

$$\sigma_x = \frac{E}{1-v^2}[\varepsilon_x + v\varepsilon_y]$$

$$\sigma_y = \frac{E}{1-v^2}[\varepsilon_y + v\varepsilon_x]$$

(8.2b)

$$\tau_{xy} = G\gamma_{xy}$$

Equation (8.2) represents a two-dimensional complex strain system (Figure 8.2) that is directly related to the two-dimensional complex stress system discussed in Section 7.1.

The collective information of ε_x, ε_y and γ_{xy} at a point within a two-dimensional system is termed as state of strain at that point.

In Figure 8.2, if the shear strain, γ_{xy}, is zero (no distortion), the two normal strains are called principal strains and designated as ε_1 and ε_2, where $\varepsilon_1 > \varepsilon_2$. The two principal strains are associated with the two principal stresses (Figure 8.3) discussed in Chapter 7.

Figure 8.2

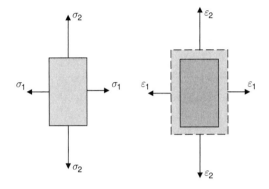

Figure 8.3

For the principal stresses and principal strains shown in Figure 8.3, Hooke's law (8.2) is reduced to the following:

- *Strains in terms of stresses:*

$$\varepsilon_1 = \frac{\sigma_1}{E} - \nu\frac{\sigma_2}{E} = \frac{1}{E}[\sigma_1 - \nu\sigma_2]$$

$$\varepsilon_2 = \frac{\sigma_2}{E} - \nu\frac{\sigma_1}{E} = \frac{1}{E}[\sigma_2 - \nu\sigma_2]$$

(8.3a)

- *Stresses in terms of strains:*

$$\sigma_1 = \frac{E}{1-\nu^2}[\varepsilon_1 + \nu\varepsilon_2]$$

$$\sigma_2 = \frac{E}{1-\nu^2}[\varepsilon_2 + \nu\varepsilon_1]$$

(8.3b)

where both shear stress and shear strain are zero.

- When a structure is subjected to external loads, the state of strain is, in general, different at different points within the structure.
- At a point in a structure, the strain in one direction is usually different to the strain in a different direction.
- State of strain (Figure 8.2) shows two normal strains and a shear strain at a point. When a normal strain is known, it means that not only the magnitude, but also the

direction of the strain are known. The shear strain (change in a right angle) is always in consistence with the shear stresses at the point.

- The maximum and minimum normal strains are called *principal strains*. They are, respectively, associated with the maximum and minimum principal stresses.
- The shear strain related to the maximum shear stress is the maximum shear strain.

8.1 Strain analysis

In most cases, the purpose of a strain analysis is to carry out a stress analysis. This is particularly true when the analysis is based on experiments. Usually, it is much easier and more straightforward to measure strains (deformation) than stresses by experiments. The measured strains can then be converted to stresses on the basis of Hooke's law. Like the stress analysis discussed in Chapter 7, the strain analysis here again deals with the following two issues:

- To determine the strain in relation to an important direction based on the measured strains aligning with a chosen co-ordinate system;
- To determine the maximum normal strain or maximum shear strain at a point. These strains may not necessarily align with the chosen coordinate directions.

(a) *Strains transformation*
The two states of strain are taken at the same point of a material (Figure 8.4). By comparing Figure 8.4 with Figure 7.5, the strains in the x–y and the x'–y' coordinate systems should have a similar relationship as that for stresses. Considering the fact that the shear strain, γ_{xy}, is related to a pair of shear stresses, τ_{xy} and τ_{yx}, only $\gamma_{xy}/2$ will appear in the equation as an equivalent to the shear stress τ_{xy}. Thus:

$$\varepsilon_\theta = \varepsilon_{x'} = \varepsilon_x \cos^2\theta + \varepsilon_y \sin^2\theta - \frac{\gamma_{xy}}{2}\sin 2\theta$$

$$= \frac{\varepsilon_x + \varepsilon_y}{2} + \frac{\varepsilon_x - \varepsilon_y}{2}\cos 2\theta - \frac{\gamma_{xy}}{2}\sin 2\theta$$

$$\frac{\gamma_\theta}{2} = \frac{\gamma_{x'y'}}{2} = \left(\frac{\varepsilon_x - \varepsilon_y}{2}\right)\sin 2\theta + \frac{\gamma_{xy}}{2}\cos 2\theta$$

> Can be obtained by replacing σ_x, σ_y and τ_{xy} with ε_x, ε_y and $\gamma_{xy}/2$ respectively, in Equation (7.2)

(8.4)

or

$$\gamma_\theta = \gamma_{x'y'} = (\varepsilon_x - \varepsilon_y)\sin 2\theta + \gamma_{xy}\cos 2\theta$$

(a)

(b)

Figure 8.4

where an anticlockwise angle from the x-axis is defined as positive.

(b) *Principal strains and their directions*

Following the same argument for strain transformation, the principal strains and their directions can be deducted from the principal stress equations (Equations (7.2) and (7.3)):

$$\varepsilon_1 = \frac{1}{2}[(\varepsilon_x + \varepsilon_y) + \sqrt{(\varepsilon_x - \varepsilon_y)^2 + \gamma_{xy}^2}]$$

$$\varepsilon_2 = \frac{1}{2}[(\varepsilon_x + \varepsilon_y) - \sqrt{(\varepsilon_x - \varepsilon_y)^2 + \gamma_{xy}^2}]$$

(8.5)

> Can be obtained by replacing σ_x, σ_y and τ_{xy} with ε_x, ε_x and $\gamma_{xy}/2$, respectively, in Equation (7.2)

$$\tan 2\theta = \frac{-\gamma_{xy}}{\varepsilon_x - \varepsilon_y}$$

$$\theta = \frac{1}{2}\tan^{-1}(\frac{-\gamma_{xy}}{\varepsilon_x - \varepsilon_y})$$

(c) *Maximum shear strain*

Again with the same argument, from Equation (7.4):

$$\frac{\gamma_{max}}{2} = \frac{\varepsilon_1 - \varepsilon_2}{2}$$

or

$$\gamma_{max} = \varepsilon_1 - \varepsilon_2$$

$$\frac{\gamma_{max}}{2} = \frac{1}{2}\sqrt{(\varepsilon_x - \varepsilon_y)^2 + \gamma_{xy}^2}$$

(8.6)

or

$$\gamma_{max} = \sqrt{(\varepsilon_x - \varepsilon_y)^2 + \gamma_{xy}^2}$$

8.2 Strain measurement by strain gauges

Experimental strain and stress analysis is an important tool in structural testing and design, where deformations or strains are usually measured and then converted to stresses. Strain gauges are by far the most commonly adopted method of measuring strains.

When a strain gauge is mounted on the surface of a structural member subject to deformation, the gauge deforms with the structure. The electrical resistance of the gauge changes as the gauge deforms. This change is recorded by the strain meter and subsequently converted to the real normal strain in the longitudinal direction of the gauge at the point of measurement (Figure 8.5).

Figure 8.5

Figure 8.6

- If the normal strain in a particular direction at a surface point of a material is required, a single strain gauge is mounted along the required direction at the point.
- If the directions of principal stresses at a surface point are known in advance, two independent strain measurements in the directions of the principal stresses are needed to obtain the principal strains and stresses.
- If the directions of principal stresses at a surface point are unknown in advance, a strain gauge rosette, which is an arrangement of three closely positioned gauge grids and separately oriented, is needed to measure the normal strains along the three different directions that are required to determine the principal strains and stresses.

The strain gauge rosette shown in Figure 8.6 is the most commonly used type of rosette and is called *rectangular rosette*.

When a rectangular rosette is mounted on the surface point of a material, the strains along the a, b and c directions of the material at the point are measured as ε_a, ε_b and ε_c, respectively. These measurements are then introduced into the following equations to determine the principal strains:

$$\varepsilon_1 = \frac{1}{2}(\varepsilon_a + \varepsilon_c) + \frac{1}{\sqrt{2}}\sqrt{(\varepsilon_a - \varepsilon_b)^2 + (\varepsilon_c - \varepsilon_b)^2}$$

$$\varepsilon_2 = \frac{1}{2}(\varepsilon_a + \varepsilon_c) - \frac{1}{\sqrt{2}}\sqrt{(\varepsilon_a - \varepsilon_b)^2 + (\varepsilon_c - \varepsilon_b)^2}$$

(8.7a)

The directions of the principal strains can be determined by:

$$\tan 2\theta = \frac{2\varepsilon_b - \varepsilon_a - \varepsilon_c}{\varepsilon_a - \varepsilon_c}$$

(8.7b)

The principal stresses at the point are found by introducing the principal strains of Equation (8.7a) into Equation (8.3b).

8.3 Key points review

8.3.1 Complex strain system

- At a point in a material, there are six independent strain components, including three normal strains and three shear strains.
- In a two-dimensional case, there are three independent strains, two normal strains and one shear strain at a point of a material.

- A strain usually varies from point to point.
- Strains are dimensionless.
- Principal strains are normal strains and include both maximum and minimum strains at a point.
- In a three-dimensional system, there are three principal strains; while in a two-dimensional system, there are two principal strains.
- Principal strains and principal stresses have the same directions, and are related by Hooke's law.

8.3.2   Strain measurement by strain gauges

- Strain gauges can only measure local strains at a point.
- Strain gauges measure the strains in the plane of the gauge.
- A strain gauge measures the normal strain in the longitudinal direction of the gauge.
- If the directions of principal stresses are known, only two strain gauges are required to determine the principal strains at a point.
- A strain gauge rosette consisting of at least three gauges is sufficient to measure the principal strains at any surface point of a material.

8.4   Examples

EXAMPLE 8.1
A thin-walled cylinder is subjected to an internal pressure as shown in Figure E8.1. At a point on the outside surface of the generator, the two strain gauges recorded are, respectively, $\varepsilon_a = 254 \times 10^{-6}$ and $\varepsilon_b = 68 \times 10^{-6}$. Determine the principal stresses at the surface point. The Young's modulus and Poisson's ratio of the cylinder are, respectively, $E = 210\,\text{GPa}$ and $\nu = 0.28$.

Figure E8.1

[Solution] *Due to the symmetric deformation and absence of shear stress on the cross-section, the state of stress can be defined by the axial and hoop stresses caused by the internal pressure.*

The two stresses are the principal stresses. The two strain measurements from the strain gauge are therefore the two principal strains. Thus a straightforward application of Equation (8.3b) yields the principal stresses at the point.

Let $\varepsilon_1 = \varepsilon_a$ and $\varepsilon_2 = \varepsilon_b$

$$\sigma_1 = \frac{E}{1 - \nu^2}[\varepsilon_1 + \nu\varepsilon_2] = \frac{210\,\text{GPa}}{1 - 0.28^2}[254 \times 10^{-6} + 0.28 \times 68 \times 10^{-6}]$$

$$= 0.062\,\text{GPa} = 62\,\text{MPa}$$

$$\sigma_2 = \frac{E}{1 - \nu^2}[\varepsilon_2 + \nu\varepsilon_1] = \frac{210\,\text{GPa}}{1 - 0.28^2}[68 \times 10^{-6} + 0.28 \times 254 \times 10^{-6}]$$

$$= 0.032\,\text{GPa} = 32\,\text{MPa}$$

The two principal stresses are, respectively, 62 MPa and 32 MPa.

EXAMPLE 8.2

At a surface point of a beam, a rectangular strain gauge rosette (Figure 8.6) recorded the following strains:

$$\varepsilon_a = -270 \times 10^{-6}, \quad \varepsilon_b = -550 \times 10^{-6}, \quad \varepsilon_c = 80 \times 10^{-6}$$

If the gauges 'a' and 'c' are in line with and perpendicular to the axis of the beam, calculate the principal stresses at the point and their direction. Take $E = 200\,\text{GPa}$ and $\nu = 0.3$.

[Solution] *This question requires direct application of Equation (8.7) to compute principal strains and their directions, and Equation (8.3b) to compute principal stresses. For isotropic materials, the directions of principal stresses coincide with the directions of principal strains.*

From Equation (8.7a):

$$\varepsilon_1 = \frac{1}{2}(\varepsilon_a + \varepsilon_c) + \frac{1}{\sqrt{2}}\sqrt{(\varepsilon_a - \varepsilon_b)^2 + (\varepsilon_c - \varepsilon_b)^2}$$

$$= \frac{1}{2}(-270 + 80) \times 10^{-6} + \frac{10^{-6}}{\sqrt{2}}\sqrt{(-270 + 550)^2 + (80 + 550)^2}$$

$$= 392.49 \times 10^{-6}$$

$$\varepsilon_2 = \frac{1}{2}(\varepsilon_a + \varepsilon_c) - \frac{1}{\sqrt{2}}\sqrt{(\varepsilon_a - \varepsilon_b)^2 + (\varepsilon_c - \varepsilon_b)^2}$$

$$= \frac{1}{2}(-270 + 80) \times 10^{-6} - \frac{10^{-6}}{\sqrt{2}}\sqrt{(-270 + 550)^2 + (80 + 550)^2}$$

$$= -582.49 \times 10^{-6}$$

From Equation (8.3b):

$$\sigma_1 = \frac{E}{1-\nu^2}[\varepsilon_1 + \nu\varepsilon_2] = \frac{200 \times 10^9 \, \text{Pa}}{1-0.3^2}[392.49 \times 10^{-6} + 0.3 \times (-582.49 \times 10^{-6})]$$

$$= 47.86 \, \text{MPa}$$

$$\sigma_2 = \frac{E}{1-\nu^2}[\varepsilon_2 + \nu\varepsilon_1] = \frac{200 \times 10^9 \, \text{Pa}}{1-0.3^2}[-582.49 \times 10^{-6} + 0.3 \times 392.49 \times 10^{-6}]$$

$$= -102.14 \, \text{MPa}$$

From Equation (8.7b):

$$\tan 2\theta = \frac{2\varepsilon_b - \varepsilon_a - \varepsilon_c}{\varepsilon_a - \varepsilon_c} = \frac{2 \times (-550) - (-270) - 80}{-270 - 80}$$

$$= 2.6$$

Thus:

$$2\theta = 68.8°$$

$$\theta = 34.4°$$

In order to determine the directions of the principal stresses (strains), the conceptual analysis below can be followed. We know that one of the principal stresses (strains) is in the direction of 34.4° from gauge 'a', and the normal strain in this direction must be compressive (negative), judging by the compressive strains of gauges 'a' and 'b'. The principal stress (strain) in this direction is the minimum principal stress (strain). Thus:

Figure E8.2

$$\sigma_1 = 47.86 \, \text{MPa}; \qquad \theta_1 = \theta + 90° = 34.4° + 90° = 124.4°$$
$$\sigma_2 = -102.14 \, \text{MPa}; \quad \theta_2 = \theta = 34.4°$$

EXAMPLE 8.3
At a point on the surface of the generator (Figure E8.3), the readings of the rectangular strain gauge rosette are $\varepsilon_a = 100 \times 10^{-6}$, $\varepsilon_b = -200 \times 10^{-6}$ and $\varepsilon_c = -300 \times 10^{-6}$. The gauges 'a' and 'c' are in line with and perpendicular to the axis of the beam. Determine the principal stresses at the surface point and the magnitudes of the applied twist moment and the axial force. The Young's modulus and Poisson's ratio of the cylinder are, respectively, $E = 70 \, \text{GPa}$ and $\nu = 0.3$.

(Continued)

EXAMPLE 8.3 *(Continued)*

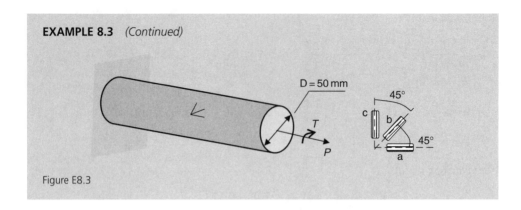

Figure E8.3

[Solution] *The principal stresses can be easily computed by Equations (8.7a) and (8.3b). To determine the applied axial force, the relationship between the principal stresses and the normal stresses on the cross-section at the point must be established, from which the normal stress and then the axial force can be calculated. To determine the applied torque, the relationship between the principal stresses and the shear stresses on the cross-section is required since the shear stress is directly related to the applied torque.*

(a) *Principal strains* (Equation 8.7a):

$$\varepsilon_1 = \frac{1}{2}[(\varepsilon_a + \varepsilon_c) + \frac{1}{\sqrt{2}}\sqrt{(\varepsilon_a - \varepsilon_b)^2 + (\varepsilon_c - \varepsilon_b)^2}$$

$$= \frac{1}{2}[(1000 - 300) \times 10^{-6} + \frac{10^{-6}}{\sqrt{2}}\sqrt{(1000 + 200)^2 + (-300 + 200)^2}$$

$$= 1202 \times 10^{-6}$$

$$\varepsilon_2 = \frac{1}{2}[(\varepsilon_{0°} + \varepsilon_{90°}) - \frac{1}{\sqrt{2}}\sqrt{(\varepsilon_{0°} - \varepsilon_{45°})^2 + (\varepsilon_{90°} - \varepsilon_{45°})^2}$$

$$= \frac{1}{2}[(1000 - 300) \times 10^{-6} - \frac{10^{-6}}{\sqrt{2}}\sqrt{(1000 + 200)^2 + (-300 + 200)^2}$$

$$= -502 \times 10^{-6}$$

(b) *Principal stresses* (Equation 8.3b):

$$\sigma_1 = \frac{E}{1 - \nu^2}(\varepsilon_1 + \nu\varepsilon_2) = \frac{70,000}{1 - 0.3^2}(1202 - 0.3 \times 502) \times 10^{-6}$$

$$= 80.09 \, \text{N/mm}^2$$

$$\sigma_2 = \frac{E}{1 - \nu^2}(\varepsilon_2 + \nu\varepsilon_1) = \frac{70,000}{1 - 0.3^2}(-502 + 0.3 \times 1202) \times 10^{-6}$$

$$= -10.9 \, \text{N/mm}^2$$

(c) *Stresses (σ_x and τ_{xy}) on the cross-section:*

Taking the axial direction as the x direction, the state of stress at the surface point is as follows:

where the normal stress, σ_x, is related to the axial force F and the shear stress, τ_{xy}, is related to the twist moment T. In this case the hoop stress is zero, that is, $\sigma_y = 0$.

From Equation (7.2), the sum of the two principal stresses yields:

$$\sigma_1 + \sigma_2 = \sigma_x + \sigma_y \text{ (This equation holds for any two-dimensional problem)}$$

Thus:

$$\sigma_x = \sigma_1 + \sigma_2 = 80.09 - 10.9 = 70\,\text{N/mm}^2 \ (\sigma_y = 0)$$

On the cross-section (Equation 2.1(a)):

$$\sigma_x = \frac{F}{A}$$

$$F = \sigma_x A = 70 \times \frac{\pi}{4} \times 50^2 = 137.4\,\text{kN}$$

Again from Equation (7.2), the difference between the two principal stresses is:

$$\sigma_1 - \sigma_2 = \sqrt{(\sigma_x - \sigma_y)^2 + 4\tau_{xy}^2}$$

In this case, $\sigma_y = 0$. Thus:

$$\tau_{xy} = \frac{1}{2}\sqrt{(\sigma_1 - \sigma_2)^2 - \sigma_x^2} = \frac{1}{2}\sqrt{(80.09 - 10.9)^2 - 70^2}$$
$$= 19.7\,\text{N/mm}^2$$

On the cross-section (Equation (3.2)):

$$\tau_{xy} = \frac{Tr}{J}$$

$$T = \frac{\tau_{xy} J}{r} = \frac{29.7 \times (\pi/32) \times 50^4}{25} = 0.7\,\text{kN m}$$

The axial force and twist moment applied at the free end are, respectively, 137.4 kN and 0.7 kN m.

EXAMPLE 8.4
A cantilever beam of solid circular cross-section has a diameter 100 mm and is 1 m long. An arm of length d is attached to its free end. The arm carries a load W as shown in Figure E8.4. On the beam's upper surface, halfway along its length and coinciding with its vertical plane of symmetry is a rectangular strain gauge rosette which gave the following readings for particular values of W and d:

$$\varepsilon_a = 1500 \times 10^{-6}, \quad \varepsilon_b = -300 \times 10^{-6}, \quad \varepsilon_c = -450 \times 10^{-6}.$$

If gauges 'a' and 'c' are in line with and perpendicular to the axis of the beam calculate the values of W and d. Take $E = 200\,000\,\text{N/mm}^2$ and $\nu = 0.3$.

Figure E8.4

[Solution] *Again, a direct application of Equations (8.7a) and (8.3b) can provide the principal strains and principal stresses at the point. To determinate the load W and then lever arm d, the bending moment and twist moment due to W and d on the cross-section at the location where the strain gauge rosette is mounted must be calculated. The relationships between the bending and twist moments and the normal and shear stresses on the section are then established, from which the values of W and d are finally determined.*

(a) *Principal strains*
Substituting values of ε_a, ε_b and ε_c in Equation (8.7a) gives:

$$\varepsilon_1 = \frac{1}{2}[(\varepsilon_a + \varepsilon_c) + \frac{1}{\sqrt{2}}\sqrt{(\varepsilon_a - \varepsilon_b)^2 + (\varepsilon_c - \varepsilon_b)^2}$$

$$= \frac{1}{2}[(1500 - 450) \times 10^{-6} + \frac{10^{-6}}{\sqrt{2}}\sqrt{(1500 + 300)^2 + (-450 + 300)^2}$$

$$= 1802.1 \times 10^{-6}$$

$$\varepsilon_2 = \frac{1}{2}[(\varepsilon_a + \varepsilon_c) - \frac{1}{\sqrt{2}}\sqrt{(\varepsilon_a - \varepsilon_b)^2 + (\varepsilon_c - \varepsilon_b)^2}$$

$$= \frac{1}{2}[(1500 - 450) \times 10^{-6} - \frac{10^{-6}}{\sqrt{2}}\sqrt{(1500 + 300)^2 + (-450 + 300)^2}$$

$$= -752.2 \times 10^{-6}$$

(b) *Principal stresses*
Now from Equation (8.3b):

$$\sigma_1 = \frac{E}{1-v^2}(\varepsilon_1 + v\varepsilon_2)$$

$$= \frac{200\,000}{1-0.3^2}(1802.2 - 0.3 \times 752.2) \times 10^{-6}$$

$$= 346.5\,\text{N/mm}^2$$

$$\sigma_2 = \frac{E}{1-v^2}(\varepsilon_2 + v\varepsilon_1)$$

$$= \frac{200\,000}{1-0.3^2}(-752.2 + 0.3 \times 1802.2) \times 10^{-6}$$

$$= -46.5\,\text{N/mm}^2$$

(c) *Stresses (σ_x and τ_{xy}) on the cross-section*
Taking the axial direction as the x direction, the state of stress at the surface point is as follows:

where the normal stress, σ_x, is related to the bending moment caused by W, and the shear stress, τ_{xy}, is related to the twist moment caused by W and d. The hoop stress is zero, that is, $\sigma_y = 0$.
From Equation (7.2), the sum of the two principal stresses yields:

$$\sigma_1 + \sigma_2 = \sigma_x + \sigma_y$$

Thus:

$$\sigma_x = \sigma_1 + \sigma_2 = 346.5 - 46.5 = 300\,\text{N/mm}^2$$

On the cross-section the bending moment is:

$$M = W \times 0.5\,\text{m}$$

From Equation (5.1a):

$$\sigma_x = \frac{My}{I} = \frac{W \times 500\,\text{mm} \times 50\,\text{mm}}{\pi \times 100^4/64\,\text{mm}^4} = 300\,\text{N/mm}^2$$

$$W = 58.9\,\text{kN}$$

Again from Equation (7.2), the difference between the two principal stresses is:

$$\sigma_1 - \sigma_2 = \sqrt{(\sigma_x - \sigma_y)^2 + 4\tau_{xy}^2}$$

$$\tau_{xy} = \frac{1}{2}\sqrt{(\sigma_1 - \sigma_2)^2 - \sigma_x^2} = \frac{1}{2}\sqrt{(346.5 + 46.5)^2 - 300^2}$$

$$= 126.9 \, \text{N/mm}^2$$

On the cross-section, the twist moment is:

$$T = W \times d = 58.9 \times d$$

From Equation (3.2):

$$\tau_{xy} = \frac{Tr}{J} = \frac{W \times d \times r}{J} = \frac{58.9 \times d \times 50}{\pi \times 100^4/32}$$

$$= 126.9 \, \text{N/mm}^2$$

$$d = 423 \, \text{mm}$$

The load applied at the arm and the distance are, respectively, 58.9 kN and 423 mm.

8.5 Conceptual questions

1. What is meant by 'state of strain' and why is it important in stress analysis?
2. What is meant by 'principal strains' and what is the importance of them?
3. How many principal strains are there at a point in a two-dimensional stress system?
4. What strain can be measured by a strain gauge?
5. At a surface point of a material if the directions of two principal stresses are known, how many strain gauges are needed to measure the principal strains? Describe how the principal stresses can be calculated.
6. At a surface point of a material if the directions of two principal stresses are unknown, how many strain gauges are needed to measure the principal strains? Describe how the principal strains and principal stresses can be calculated.
7. Discuss how can the Young's modulus and Poisson's ratio be measured by the strain gauges technique.
8. Can strain gauges be used to measure shear strain? Explain how this can be done.

8.6 Mini test

Problem 8.1:   A bar of circular section is subjected to torsion only. A single strain gauge is used to determine the shear stress on the cross-section. What is the best orientation in which the strain gauge should be mounted on the bar?

Figure P8.1

Problem 8.2:   In the simple tension test shown in the figure, if the two strain gauges recorded ε_a and ε_b, respectively, describe how the Young's modulus and Poisson's ratio can be determined.

Figure P8.2

Problem 8.3:   The single strain gauge on the top surface of a cantilever of rectangular section recorded a longitudinal strain ε_0. If load W is applied to the free end of the cantilever, determine the longitudinal stress at the location and the applied load W. The Young's modulus of the material is E.

Figure P8.3

Problem 8.4:   The strain readings from a rectangular strain gauge rosette are given below:

$$\varepsilon_a = 1000 \times 10^{-6}, \quad \varepsilon_b = 200 \times 10^{-6}, \quad \varepsilon_c = -600 \times 10^{-6}$$

which are the strains recorded in the 0°, 45° and 90° directions, respectively. Find the principal strains and their orientations. Find also the maximum shear strain.

Problem 8.5:   The cantilever column of rectangular cross-section shown in Figure 8.5 is subjected to a wind pressure of intensity ω and a compressive force F. A rectangular strain gauge rosette located at a surface point Q is positioned on the centroidal axis 2 m downwards from the free end. The rosette recorded the following strain readings:

$$\varepsilon_a = -222 \times 10^{-6}, \varepsilon_b = -213 \times 10^{-6}, \varepsilon_c = +45 \times 10^{-6},$$

where gauges 'a', 'b' and 'c' are in line with, at 45° to and perpendicular to the centroidal axis of the column, respectively. Calculate the direct stress and the shear stress at point Q in the horizontal plane and hence the compressive force F and the transverse pressure ω. $E = 31,000 \,\text{N/mm}^2$, $\nu = 0.2$.

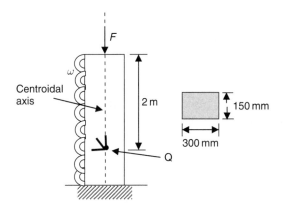

Figure P8.5

9 Theories of elastic failure

When a structural component is subjected to increasing loads it eventually fails. Failure is a condition that prevents a structure from performing the intended task.

In practical applications, failure can be defined as:

- Fracture with very little yielding
- Permanent deformation

The resistance to failure of a material is called *strength*. It is comparatively easy to determine the strength or the point of failure of a component subjected to a single tensile force. For example, for the bars shown in Figure 9.1(a), the material fractures when the principal stress approaches the fracture stress in a tensile test. This failure mode occurs normally for bars made of brittle materials, such as cast iron, and is best demonstrated by such a bar subjected to torsion (Figure 9.1(b)), where the maximum principal stress acts in the direction 45° to the longitudinal axis.

However, if the material of the bars shown in Figure 9.1 is replaced by a ductile material, for example, mild steel, the failure mode is significantly different. For the bar in tension (Figure 9.2a), the bar breaks after undergoing permanent local deformation (yielding). Compared with the bar subjected to torsion in Figure 9.1, the failure surface of the ductile bar is almost normal to the longitudinal axis (Figure 9.2b), where the shear stress has reached the shear strength of the material.

From the above simple examples, it can be concluded that material property is a predominant factor that must be considered in failure analysis.

When a material is subjected to a combination of tensile, compressive and shear stresses, it is far more complicated to determine whether or not the material has failed, and how the material will fail. Most of the information on yielding or fracture of material subjected to complex stress system comes from practical design experience, experimental evidence and interpretation of them. Investigations on the information enable a formulation of theories of failure to be established for various materials subjected to complex stresses.

The establishment of failure criteria for complex stress system is based on the extension of the concept of failure criteria for a material subjected to a uniaxial stress to materials subjected to combined stresses. The basis for this extension is the introduction of an equivalent stress (σ_{eq}) that represents a combined action of the stress components of a complex stress system. It is assumed that failure will occur in a material subjected to a complex stress system when this

Figure 9.1

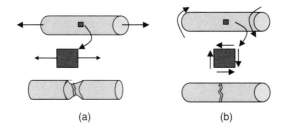

Figure 9.2

equivalent stress reaches a limiting value (σ_{Yield}) that is equal to the failure (yield or fracture) stress of the same material subjected to simple tension.

Materials can be broadly separated into ductile and brittle materials. Examples of ductile materials include mild steel, copper, etc. Cast iron and concrete are typical examples of brittle materials. Brittle materials experience little deformation prior to failure and failure is generally sudden. A ductile material is considered to have failed when a marked plastic deformation has begun. A number of theories of elastic failure are recognized, including the following:

- Maximum principal stress theory
- Maximum shear stress theory (Tresca theory)
- Maximum distortional energy density theory (von Mises theory).

The selection of failure criteria usually depends on a number of aspects of a particular design, including material properties, state of stress, temperature and design philosophy. It may be possible that there exist several failure criteria that are applicable to a material in a particular design. However, in most cases, failure criteria are classified as applicable to brittle or ductile materials.

9.1 Maximum principal stress criterion

This criterion assumes that principal stress is the driving factor that causes failure of materials. According to this criterion, the following comparison has been made between a complex stress system and a simple tension/compression test.

Two-dimensional complex state of stress	Simple tension/compression *at failure*
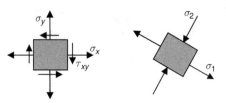 From Equation (7.2) the principal stresses are: $$\sigma_1 = \frac{1}{2}[(\sigma_x + \sigma_y) + \sqrt{(\sigma_x - \sigma_y)^2 + 4\tau_{xy}^2}]$$ $$\sigma_2 = \frac{1}{2}[(\sigma_x + \sigma_y) - \sqrt{(\sigma_x - \sigma_y)^2 + 4\tau_{xy}^2}]$$	Tension Compression

At failure the stresses in the two dimensional complex system are:

(i)   when $\sigma_2 \geq 0$:

   $\sigma_{eq} = \sigma_1 = \sigma_{Yield}$ (in tension)

(ii)   when $\sigma_1 \leq 0$:

   $\sigma_{eq} = |\sigma_2| = \sigma_{Yield}$ (in compression)   (9.1)

(iii)   when $\sigma_1 > 0$, $\sigma_2 < 0$:

   $\sigma_{eq} = \sigma_1 = \sigma_{Yield}$ (in tension)

   and

   $\sigma_{eq} = |\sigma_2| = \sigma_{Yield}$ (in compression)

From the above comparison, it is concluded that for an arbitrary state of stress

Failure (i.e. yielding) will occur when one of the principal stresses in a material is equal to the yield stress in the same material at failure in simple tension or compression.

The criterion was intended to work for brittle and ductile materials, while experimental evidence shows that it is approximately correct only for brittle materials. For brittle materials, both tensile and compressive strength should be checked upon since they are usually different.

9.2   Maximum shear stress criterion (Tresca theory)

This criterion assumes that shear stress is the driving factor that causes failure of a material. According to this criterion, comparisons are made between the maximum shear stress of a material subjected to a complex stress system and that of the same material at failure when a simple tension is applied.

Two-dimensional complex state of stress	Simple tension *at failure*
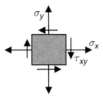 From Equation (7.4) the maximum shear stress is: $$\tau_{max} = \frac{1}{2}\sqrt{(\sigma_x - \sigma_y)^2 + 4\tau_{xy}^2}$$ or $$\tau_{max} = \frac{\sigma_1 - \sigma_2}{2}$$	Since there is no shear stress, $\sigma_x (= \sigma_{Yield})$ and $\sigma_y (= 0)$ are the two principal stresses. At the moment of failure the maximum shear stress of the above state of stress is: $$\tau_{max} = \frac{\sigma_1 - \sigma_2}{2} = \frac{\sigma_x}{2} = \frac{\sigma_{Yield}}{2}$$

At failure the stresses in the two-dimensional complex system are:

$$\sigma_{eq} = \sqrt{(\sigma_x - \sigma_y)^2 + 4\tau_{xy}^2} = \sigma_{Yield} \qquad\qquad (9.2)$$

or

$$\sigma_{eq} = \sigma_1 - \sigma_2 = \sigma_{Yield}$$

From the above comparison, it is concluded that for an arbitrary state of stress

Failure (i.e. yielding) will occur when the maximum shear stress in the material is equal to the maximum shear stress in the same material at failure in simple tension.

9.3   Distortional energy density (von Mises theory) criterion

This criterion assumes that distortional energy density (shear strain energy per unit volume) is the driving factor that causes failure of a material. According to this criterion, comparisons are made between the maximum shear strain energy per unit volume of a material subjected to a complex state of stress and that of the same material at failure when a simple tension is applied. Thus,

Two-dimensional complex stress system	Simple tension *at failure*
The shear strain energy per unit volume is:	Since there is no shear stress, $\sigma_x(=\sigma_{Yield})$ and $\sigma_y(=0)$ are the two principal stresses. At the moment of failure the shear strain energy per unit volume is:
$$U_s = \frac{1}{12G}[(\sigma_1 - \sigma_2)^2 + \sigma_2^2 + \sigma_1^2]$$ $$= \frac{1}{6G}[\sigma_1^2 + \sigma_2^2 - \sigma_1\sigma_2]$$ or $$U_s = \frac{1}{6G}[\sigma_x^2 - \sigma_x\sigma_y + \sigma_y^2 + 3\tau_{xy}^2]$$	$$U_s = \frac{1}{6G}[\sigma_1^2 + \sigma_2^2 - \sigma_1\sigma_2]$$ $$= \frac{1}{6G}\sigma_{Yield}^2$$

At failure the stresses in the two-dimensional complex system are:

$$\sigma_{eq} = \sqrt{\sigma_1^2 + \sigma_2^2 - \sigma_1\sigma_2} = \sigma_{Yield} \qquad (9.3)$$

$$\sigma_{eq} = \sqrt{\sigma_x^2 - \sigma_x\sigma_y + \sigma_y^2 + 3\tau_{xy}^2} = \sigma_{Yield}$$

From the above comparison, it is concluded that for an arbitrary state of stress

Failure (i.e. yielding) will occur when the shear strain energy per unit volume in a material is equal to the equivalent value at failure of the same material in simple tension.

The application of the failure criterions depends on the modes of failure (e.g. failure by yielding or fracture). In general, the maximum principal stress criterion is valid for failure mode dominated by fracture in brittle materials, while Tresca and von Mises criterions are valid for general yielding mode of failure in ductile materials.

9.4 Special forms of Tresca and von Mises criterions

In many practical applications, at a surface point of a material there exists normal stress in one direction only. For example, at a surface point of the beam shown in Figure 9.3, the normal stress in the vertical direction is far smaller than that in the longitudinal direction and is usually ignored in the stress analysis of beams. If the longitudinal direction is defined as the x direction, $\sigma_y = 0$. Thus, Tresca and von Mises theories (Equations (9.2) and (9.3), respectively) are reduced to the following:

Figure 9.3

Tresca theory:

$$\sqrt{\sigma_x + 4\tau_{xy}^2} = \sigma_{\text{Yield}} \tag{9.4a}$$

von Mises theory:

$$\sqrt{\sigma_x^2 + 3\tau_{xy}^2} = \sigma_{\text{Yield}} \tag{9.4b}$$

It is clear from Equation (9.4) that Tresca theory is more conservative than von Mises theory since the shear stress is factorized by 4 rather than 3.

9.5   Key points review

- A brittle material is likely to fail by fracture, and thus has higher compressive strength (holds greater compressive loads).
- A ductile material is likely to fail by yielding, i.e., having permanent deformation.
- The maximum principal stress theory is best for brittle materials and can be unsafe for ductile materials.
- The maximum shear stress and the distortional energy density theories are suitable for ductile materials, while the former is more conservative than the latter.
- For brittle materials having a weaker tensile strength, reinforcement is usually required in the tension zone to increase the load-carrying capacity.

9.6   Recommended procedure of solution

It is possible that a material may fail at any point within the material, but, in general, starting at a point where an equivalent stress defined above reaches a critical value first. Therefore, in a practical design, the application of the above criterions relies on identification of, for example, in the design of a beam, critical cross-sections where maximum bending moment, twist moment or axial force may exist. On these critical sections, maximum normal and shear stresses are found. A recommended procedure of solution is shown in Figure 9.4.

9.7   Examples

EXAMPLE 9.1
Explain why concrete is normally reinforced with steel bars or rods when tensile forces are applied to a structure.

[Solution] *This question tests your understanding of the failure mode of brittle materials. This type of materials usually has different strength in tension and compression.*

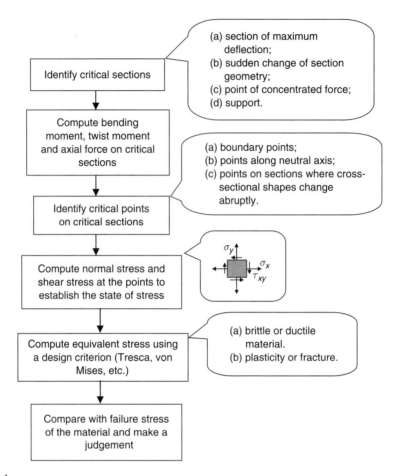

Figure 9.4

Concrete is a typical example of brittle material that is weaker in tension, and has higher load capacity in compression. When concrete is subjected to tension, fracture is initiated at imperfections or micro-cracks, whereas the imperfections and micro-cracks are closed in compression and fracture is unlikely to occur. Therefore, steel or other types of reinforcements are needed in the tension zone to increase the tensile strength of the structure to prevent early fracture failure.

EXAMPLE 9.2

Codes of practice for the use of structural steel uses either Tresca or von Mises criterion. For a beam member subjected to bending and shear, the criterions can be expressed as

Tresca:

$$\sqrt{\sigma_x^2 + 4\tau_{xy}^2} = \sigma_{\text{Yield}}$$

Von Mises:

$$\sqrt{\sigma_x^2 + 3\tau_{xy}^2} = \sigma_{\text{Yield}}$$

Verify the expression and state which criterion is more conservative.

[Solution] *The answer to this question is a direct application of Equations (9.2) and (9.3) to the state of stress where one of the normal stresses is zero.*

For the beam shown in Figure E9.2(a) subjected to bending and shearing, the state of stress at the arbitrary point is represented by Figure E9.2(b), where only one normal stress exists.

(a)   (b)

Figure E9.2

Assuming that the horizontal and vertical directions are, respectively, the x and y directions, $\sigma_y = 0$. Thus the two principal stresses at the point are:

$$\sigma_1 = \frac{1}{2}[\sigma_x + \sqrt{(\sigma_x)^2 + 4\tau_{xy}^2}]$$

$$\sigma_2 = \frac{1}{2}[\sigma_x - \sqrt{(\sigma_x)^2 + 4\tau_{xy}^2}]$$

Introducing the obtained principal stresses in to Equations (9.2) and (9.3) yields, respectively, the special form of the two criterions.

Since $\sqrt{\sigma_x^2 + 4\tau_{xy}^2} > \sqrt{\sigma_x^2 + 3\tau_{xy}^2}$, for the same value of σ_{Yield} a design by Tresca criterion demands smaller σ_x or/and τ_{xy}, that is, a reduction of applied loads or an increase of material usage. Thus Tresca criterion is more conservative than von Mises criterion.

EXAMPLE 9.3

Consider a bar of cast iron under complex loading. The bar is subjected to a bending moment of $M = 39\,\mathrm{N\,m}$ and a twist moment of $T = 225\,\mathrm{N\,m}$. The diameter of the bar is $D = 20\,\mathrm{mm}$. If the material of the bar fails at $\sigma_{Yield} = 128\,\mathrm{MPa}$ in a simple tension test, will failure of the bar occur according to the maximum principal stress criterion?

Figure E9.3

[Solution] *On any cross-section of the beam, the maximum shear stress due to twisting is along the circumference and the maximum tensile stress due to bending is at the lowest edge. Thus, the state of stress at a point A taken from the lowest generator must be considered. The stresses at this point can be calculated from Equations (3.2) and (5.1a). The principal stresses can then be calculated from these stresses.*

The shear stress due to torsion is:

$$\tau_{xy} = \frac{Tr}{J} = \frac{TD/2}{\pi D^4/32} = \frac{32 \times 225\,\text{N m} \times 20 \times 10^{-2}\text{m}/2}{\pi \times (20 \times 10^{-3})^4}$$

$$= 143.24 \times 10^6 \, N/m^2 = 143.24 \, MPa$$

The normal stress due to bending is:

$$\sigma_x = \frac{My}{I} = \frac{MD/2}{\pi D^4/64} = \frac{64 \times 39\,\text{N m} \times 20 \times 10^{-3}\,\text{m}/2}{\pi \times (20 \times 10^{-3})^4}$$

$$= 49.66 \times 10^6 \, N/m^2 = 49.66 \, MPa$$

From Equation (7.2), the principal stresses at the point are:

$$\sigma_1 = \frac{1}{2}[(\sigma_x + \sigma_y) + \sqrt{(\sigma_x - \sigma_y)^2 + 4\tau_{xy}^2}]$$

$$= \frac{1}{2}[(49.66\,\text{MPa} + 0) + \sqrt{(49.66\,\text{MPa} - 0)^2 + 4 \times (143.24\,\text{MPa})^2}]$$

$$= 169.96 \, MPa$$

$$\sigma_2 = \frac{1}{2}[(\sigma_x + \sigma_y) - \sqrt{(\sigma_x - \sigma_y)^2 + 4\tau_{xy}^2}]$$

$$= \frac{1}{2}[(49.66\,\text{MPa} + 0) - \sqrt{(49.66\,\text{MPa} - 0)^2 + 4 \times (143.24\,\text{MPa})^2}]$$

$$= -120.3 \, MPa$$

$$\sigma_1 > \sigma_{Yield}$$

The material has failed according to the maximum principal stress criterion.

EXAMPLE 9.4
In a tensile test on a metal specimen having a cross-section of 20 mm × 10 mm failure occurred at a load of 70 kN. A thin plate made from the same material is subjected to loads such that at a certain point in the plate the stresses are $\sigma_y = -70\,\text{N/mm}^2$, $\tau_{xy} = 60\,\text{N/mm}^2$. Determine from the von Mises and Tresca criterions the maximum allowable tensile stress, σ_x, that can be applied at the same point.

Figure E9.4

[Solution] *For the state of stress shown in Figure E9.4, the equivalent stresses of Equations (9.2) and (9.3) can be calculated and they are functions of σ_x. Comparing the equivalent stresses with the failure stress of the material at simple tension test yields the maximum allowable σ_x.*

From the simple tension test:

$$\sigma_{Yield} = \frac{70000\,N}{20\,mm \times 10\,mm} = 350\,N/mm^2$$

Tresca criterion (Equation (9.2)):

$$\sigma_{eq} = \sqrt{(\sigma_x - \sigma_y)^2 + 4\tau_{xy}^2} \le \sigma_{Yield}$$

$$\sqrt{(\sigma_x - (-70))^2 + 4 \times 60^2} \le 350$$

Therefore:

$$(\sigma_x + 70)^2 + 4 \times 60^2 \le 350^2$$
$$\sigma_x^2 + 140x - 103200 \le 0$$

Thus,

$$\sigma_x \le 259\,N/mm^2$$

von Mises criterion (Equation (9.3)):

$$\sigma_{eq} = \sqrt{\sigma_x^2 - \sigma_x \sigma_y + \sigma_y^2 + 3\tau_{xy}^2} \le \sigma_{Yield}$$

$$\sqrt{\sigma_x^2 - \sigma_x(-70) + (-70)^2 + 3 \times 60^2} \le 350$$

Therefore:

$$\sigma_x^2 + 70\sigma_x - 106800 \le 0$$

Thus:

$$\sigma_x \le 293.7\,N/mm^2$$

From the above solutions, it can be seen that the maximum allowable value of σ_x from Tresca criterion is smaller than that from von Mises criterion, and hence the resulting design is more conservative.

9.8 Conceptual questions

1. Describe the differences between failure by fracture and failure by yielding.
2. Describe the failure mode of a cast iron bar subjected to a uniaxial tensile force. If the bar is made of mild steel, describe how the failure mode is different.
3. Describe the failure mode of a cast iron bar subjected to torsion. If the bar is made of mild steel, describe how the failure mode is different.
4. What is the definition of the maximum principal stress criterion? What type of material is it useful for?
5. What is the definition of the Tresca failure criterion? What type of material is it useful for?
6. What is the definition of the von Mises failure criterion? What type of material is it useful for?
7. Explain the term 'equivalent stress' as used in connection with failure criterions.
8. On what conditions the simplified form of Tresca and von Mises criterions can be used?

9.9 Mini test

Problem 9.1: Explain why different failure criterions are needed in design and discuss what aspects should be considered in a typical design.

Problem 9.2: In a ductile material there are four points at which the states of stress are, respectively, as follows:

Figure P9.2

Which point fails first and which point fails at last? What conclusion can you draw from your analysis?

Problem 9.3: The stresses at a point of a two-dimensional structural member are found as follows:

$$\sigma_x = 140\,\text{N/mm}^2, \sigma_y = -70\,\text{N/mm}^2, \tau_{xy} = 60\,\text{N/mm}^2$$

The material of the member has a yield stress in simple tension of $225\,\text{N/mm}^2$. Determine whether or not failure has occurred according to Tresca and von Mises criterions.

Problem 9.4: On the beam section shown in Figure P9.4, there exists an axial force of $60\,\text{kN}$. Determine the maximum shear force that can be applied to the section using the Tresca and von Mises criterions. The material of the beam breaks down at a stress of $150\,\text{N/mm}^2$ in a simple tension test.

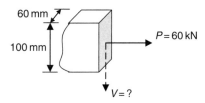

Figure P9.4

Problem 9.5:   A cantilever of circular cross-section is made from steel, which when subjected to simple tension suffers elastic breakdown at a stress of 150 N/mm². If the cantilever supports a bending moment of 25 kN m and a torque of 50 kN m, determine the minimum diameter of the cantilever using the von Mises and Tresca criterions.

Figure P9.5

10 Buckling of columns

Compression members, such as columns, are mainly subjected to axial compressive forces. The stress on a cross-section in a compression member is therefore normal compressive stress. A short column usually fails due to yielding or shearing (Figure 10.1), depending on material properties of the column.

However, when a compression member becomes longer, it could become laterally unstable and eventually collapse through sideways buckling at an axial compression. The compressive load could be far smaller than the one that would cause material failure of the same member. A simple test to illustrate this can be easily carried out by pushing the ends of a piece of spaghetti. A lateral deflection from the original position will be observed when the applied compression reaches a certain value, which is designated P_{cr} and called *critical buckling load* (Figure 10.2). The critical buckling load is the maximum load that can be applied to a column without causing instability. Any increase in the load will cause the column to fail by buckling.

Ductile   Brittle

Figure 10.1

Figure 10.2

The onset of the lateral deflection at the critical stage when $P = P_{cr}$ establishes the initiation of buckling, at which the structural system becomes unstable. P_{cr} represents the ultimate capacity of compression members, such as long columns and thin-walled structural components.

- Buckling may occur when a long or slender member is subjected to axial compression;
- A compression member may lose stability due to a compressive force that is smaller than the one causing material failure of the same member;
- The buckling failure of a compression member is related to the strength, stiffness of the material and also the geometry (slenderness ratio) of the member.

For a long or slender member subjected to compression, considering material strength along is usually not sufficient in design.

10.1   Euler formulas for columns

At the critical buckling load a column may buckle or deflect in any lateral direction. In general, the flexural rigidity or stiffness of a column is not the same in all directions (Figure 10.3). By common sense, a column will buckle (deflect) in a direction related to the minimum flexural rigidity, EI, that is, the minimum second moment of area of the cross-section.

The maximum axial load that a long, slender ideal column can carry without buckling was derived by Leonhard Euler in the eighteenth century, and termed *Euler formula* for columns. The formula was derived for ideal columns that are perfectly straight, homogeneous and free from initial stresses.

10.1.1   Euler formula for columns with pinned ends

The critical buckling load for the column shown in Figure 10.4 is:

$$P_{cr} = \frac{\pi^2 EI}{L^2} \tag{10.1}$$

For the rectangular section shown, the flexural rigidity, EI, is related to the $X–X$ axis, about which the second moment of area of the cross-section is minimum. By Equation (10.1), a higher compressive force is required to cause buckling of a shorter column or a column with greater flexural rigidity.

Equation (10.1) is the Euler formula for a column with pinned ends and is often referred to as the fundamental case. In general, columns do not always have simply supported ends. Therefore, the formula for the critical buckling load needs to be extended to include other form of end supports.

Figure 10.3

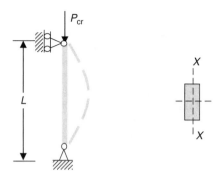

Figure 10.4

10.1.2 Euler formulas for columns with other ends

Figure 10.5 shows comparisons between columns with different end supports. Comparing with the pin-ended column (Case (a)), the buckling mode of the column with free–fixed ends (Case (b)) is equivalent to the one of a pinned column with doubled length.

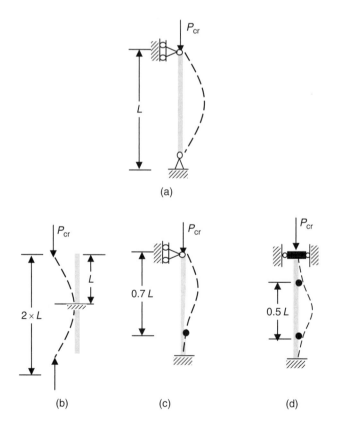

Figure 10.5

Table 10.1 Effective length factor of compression members

Support at column ends	Effective length factor λ
Fixed–Fixed	0.5
Pinned–Fixed	0.7
Pinned–Pinned	1
Free–Free	1
Fixed–Free	2

Thus for Case (b):

$$P_{cr} = \frac{\pi^2 EI}{(2L)^2} \tag{10.2}$$

For Cases (c) and (d), the Euler formulas can also be made by comparing their buckling modes with that of Case (a). In Cases (c) and (d) the deflection curves of the upper $0.7L$ and middle $0.5L$ show, respectively, a similar pattern to that shown in Case (a). The considered deflection shapes are the segments between the two inflection points on the deflection curves. The distance between the two inflection points is called *effective length*, L_e, of a column when comparison is made with a similar column with two pinned ends. For a general case, therefore, the Euler formula can be written as:

$$P_{cr} = \frac{\pi^2 EI}{L_e^2} = \frac{\pi^2 EI}{(\lambda L)^2} \tag{10.3}$$

where λ is the effective length factor that depends on the end restraints. Table 10.1 shows the values of λ for some typical end restraints.

10.2   Limitations of Euler formulas

The Euler formulas are applicable only while the material remains linearly elastic. To apply this limitation in practical designs, Equation (10.3) is rewritten as:

$$P_{cr} = \frac{\pi^2 EI}{(\lambda L)^2} = \frac{\pi^2 EAr^2}{(\lambda L)^2} = \frac{\pi^2 EA}{(\lambda L/r)^2} \tag{10.4a}$$

or, the compressive critical stress is:

$$\sigma_{cr} = \frac{P_{cr}}{A} = \frac{\pi^2 E}{(\lambda L/r)^2} \tag{10.4b}$$

In Equation (10.4) the second moment of area, I, is replaced by Ar^2, where A is the cross-sectional area and r is its radius of gyration. $\lambda L/r$ is known as *slenderness ratio*. It is clear that:

- if a column is long and slender, $\lambda L/r$ is large and σ_{cr} is small;
- if a column is short and has a large cross-sectional area, $\lambda L/r$ is small and σ_{cr} is large.

To ensure that the material of a column remains linearly elastic, the stress in the column before buckling must remain below some particular value, normally the proportionality limit of the material, σ_p . Thus:

$$\sigma_{cr} = \frac{\pi^2 E}{(\lambda L/r)^2} \leq \sigma_p \qquad\qquad (10.5a)$$

or

$$\frac{\lambda L}{r} \geq \pi \sqrt{\frac{E}{\sigma_p}} \qquad\qquad (10.5b)$$

If the slenderness ratio of a column is smaller than the specified value in Equation (10.5b), the Euler formulas are not applicable.

10.3   Key points review

- For a bar or a column subjected to compressive loading, a sudden failure (collapse) may occur. This type of failure is called *buckling*.
- A compression member may fail due to buckling at a stress level which is far below the compressive strength of the material.
- The minimum compressive load that causes buckling of a compressive member is known as critical buckling load.
- For a compression member, the critical buckling load is proportional to its flexural rigidity and inversely proportional to the square of its length.
- The end support conditions have a significant influence on the critical load, which determine the number of inflection points on the deflected column.
- The closer the inflection points of a column are, the higher the critical buckling load of the column is.
- The Euler formula for buckling of a column is based on the following assumptions:

    - The column is initially perfectly straight.
    - The compression is applied axially.
    - The column is very long in comparison with its cross-sectional dimensions.
    - The column is uniform throughout and the proportional limit is not exceeded.

- The ratio of the effective length to the radius of gyration of a column is defined as *slenderness ratio*.
- Slenderness ratio governs the critical buckling load: the larger the slenderness ratio is, the lesser the strength of a column. This means the buckling resistance decreases as the slenderness ratio increases.
- The application of Euler formulas depends on elasticity rather than compressive strength of material.
- For a particular column cross-section, length and end supports, the critical buckling load capacity depends only upon the Young's modulus E. Since there is little variation in E among different grades of steel, there is no advantage in using a high-strength steel.

10.4 Examples

EXAMPLE 10.1
Derive the Euler formula for a column with pinned ends.

Figure E10.1

[Solution] *The buckled shape shown in Figure E10.1 is possible only when the applied load is greater than the critical load. Otherwise, the column remains straight. The solution of this problem is to seek the relationship between the applied axial load and the lateral deflection of the column.*

From Equation (6.1), the deflection and the bending moment on the cross-section at an arbitrary location x is related by:

$$\frac{d^2y}{dx^2} = -\frac{M(x)}{EI} = -\frac{Py}{EI}$$

The above equation can be rewritten as:

$$\frac{d^2y}{dx^2} + k^2y = 0$$

where $k^2 = P/EI$

The equation is a linear ordinary differential equation of second order and its solution is:

$$y = A \sin kx + B \cos kx$$

where A and B are arbitrary constants that can be determined from the end support conditions. These conditions are:

at $x = 0; y(0) = 0$

at $x = L; y(L) = 0$

Hence:

$$y(0) = 0; \quad A \sin 0 + B \cos 0 = 0,$$

or

$$B = 0$$
$$y(L) = 0; \quad A \sin kL + B \cos kL = 0$$

or

$$A \sin kL = 0$$

The above condition can be satisfied by letting $A = 0$, but this will lead to no lateral deflection or buckling. Alternatively, the condition can be satisfied by taking:

$$kL = m\pi$$

where m is an nonzero integer. Thus:

$$\sqrt{\frac{P}{EI}}L = m\pi$$

or

$$P = \frac{m^2 \pi^2 EI}{L^2}$$

The above force provides a series of buckling loads at which the column will buckle with different shapes. Obviously, the lowest value of the force is the critical buckling load as defined in Figure 10.2 and takes the value when $m = 1$, that is:

$$P_{cr} = \frac{\pi^2 EI}{L^2}$$

This result is identical to the solution given in Equation (10.1).

EXAMPLE 10.2
The pin-connected plane steel truss shown in Figure E10.2 carries a concentrated force F. Assuming that both members have a circular section with a diameter of 80 mm, determine the critical buckling load of the truss ($E = 200$ GPa).

Figure E10.2

[Solution] *The critical load of the truss is the minimum load that will cause buckling of at least one member. The smallest critical buckling load of the two pin-ended members must be calculated first, to which the critical buckling load of the truss is related through equilibrium of the joint.*

For the long member:

$$P_{cr}^L = \frac{\pi^2 EI}{L^2} = \frac{\pi^2 \times 20 \times 10^9 \text{ Pa} \times \pi \times (80 \times 10^{-3})^4/64}{(4\,\text{m} \times \cos 30°)^2} = 330.7\,\text{kN}$$

For the short member:

$$P_{cr}^S = \frac{\pi^2 EI}{L^2} = \frac{\pi^2 \times 20 \times 10^9 \text{ Pa} \times \pi \times (80 \times 10^{-3})^4/64}{(4\,\text{m} \times \sin 30°)^2} = 992.2\,\text{kN}$$

So, the long member will buckle first at an axial compression of 330.7 kN. For the equilibrium of the pin joint, taking $P^L = P_{cr}^L$ and resolving in the P^L direction yield:

$$P_{cr}^L = F_{cr} \times \cos 60°$$

or

$$F_{cr} = \frac{P_{cr}^L}{\cos 60°} = \frac{330.7\,\text{kN}}{\cos 60°} = 661.4\,\text{kN}$$

When $F = F_{cr} = 661.4\,\text{kN}$, the axial compression in the short member is:

$$P^S = F \times \cos 30° = 661.4 \times \frac{\sqrt{3}}{2} = 572.8\,\text{kN}$$

The short member is stable. Thus, the critical buckling load of the truss is 661.4 kN.

EXAMPLE 10.3

A beam–column structure (Figure E10.3) consists of a beam of rectangular section (200 mm × 300 mm) and a column of I-shaped section (UBs 178 × 102 × 19). The allowable stress of the beam is $\sigma_{allow} = 100\,\text{MPa}$. The coloumn is made of steel having $E = 200\,\text{GPa}$ and $\sigma_p = 200\,\text{MPa}$. Determine the maximum point load that can be applied along the beam.

Figure E10.3

[Solution] *To determine the maximum load, both the beam and the column should be considered. The bending strength of the beam and the instability of the column must be checked. For the beam, when the force is applied at the mid-span, the maximum bending moment and the maximum normal stress occur on the mid-span section. The compression in the column is maximum when the force moves to the free end of the beam.*

For the beam, when the force is applied at the mid-span, the maximum bending moment occurs on the mid-span cross-section and is:

$$M_{max} = \frac{PL}{4} = \frac{P \times 12\,m}{4} = 3P(N\,m)$$

The maximum normal stress on the section is:

$$\sigma_{max} = \frac{M_{max}h/2}{bh^3/12} = \frac{3P \times 300 \times 10^{-3}\,m/2}{200 \times 10^{-3}\,m \times (300 \times 10^{-3}\,m)^3/12} = P \times 10^3\,N/m^2$$

From $\sigma_{max} \leq \sigma_{allow} = 100\,MPa$:

$$P \leq 100\,KN$$

For the column (UBs 178 × 102 × 19), the radius of gyration can be found as:

$$r_x = 7.48\,cm, \quad r_y = 2.37\,cm, \quad A = 24.3\,cm^2$$

The column may buckle in the plane of X–X or Y–Y, depending on the slenderness ratios related to these directions. In the X–X plane, the column can be taken as supported by pin and fixed ends and in the Y–Y plane the column can be considered as one with free–fixed ends. Thus, the slenderness ratios of the column about the X–X and Y–Y axes are, respectively (see Table 10.1):

$$\frac{\lambda_x L}{r_x} = \frac{2 \times 4\,m}{7.48 \times 10^{-2}\,m} = 107$$

$$\frac{\lambda_y L}{r_y} = \frac{0.7 \times 4\,m}{2.37 \times 10^{-2}\,m} = 118.1$$

From Equation (10.5b):

$$\frac{\lambda_x L}{r_x} = 107 > \pi\sqrt{\frac{E}{\sigma_p}} = \pi\sqrt{\frac{200 \times 10^9\,Pa}{200 \times 10^6\,Pa}} = 99.3$$

$$\frac{\lambda_y L}{r_y} = 118 > \pi\sqrt{\frac{E}{\sigma_p}} = \pi\sqrt{\frac{200 \times 10^9\,Pa}{200 \times 10^6\,Pa}} = 99.3$$

Since the slenderness ratio of the column is larger than the value specified in Equation (10.3), the Euler formulas are valid. Buckling about the Y–Y axis is more likely due to the greater slenderness ratio. Thus, from Equation (10.4a), the critical buckling load of the column is:

$$P_{cr} = \frac{\pi^2 EA}{(\lambda_y L/r_y)^2} = \frac{\pi^2 \times 200 \times 10^9\,Pa \times 24.3 \times 10^{-4}\,m^2}{(118.1)^2} = 343.9\,kN$$

The maximum compression in the column caused by the applied force occurs when the force acts at the free end of the beam. The compression can be easily calculated from the equilibrium of the beam

by taking moment about A. So:

$$12\,m \times R_B - (12+2)m \times P = 0$$

which yields:

$$R_B = \frac{7}{6}P$$

or

$$P = \frac{6}{7}R_B$$

The compression in the column is numerically equal to R_B. When R_B reaches the critical value, that is, $R_B = P_{cr}$

$$P \le \frac{6}{7}R_B = \frac{6}{7}P_{cr} = \frac{6}{7} \times 343.9\,kN = 294.8\,kN$$

Compared with the maximum force obtained from applying the strength condition of the beam, which is less than 294.8 kN, the maximum applied force that the structure can carry is therefore 100 kN.

EXAMPLE 10.4
A steel column of 4 m long is pinned at both ends and has an American Standard Steel Channel section (C130 × 10). The Young's modulus of the steel is $E = 200\,GPa$ and the elastic limit of the material is 200 MPa.

(a) Determine the critical buckling load of the column.
(b) If a column is built up of two C130 × 10 channels placed back to back at a distance of d and connected to each other at an interval of h along the length of the column, determine the values of d and h and calculate the critical buckling load of the column.

(Continued)

EXAMPLE 10.4   *(Continued)*

Figure E10.4

[Solution] *A single channel section is not symmetric. Such a column will buckle in the direction of minimum radius of gyration. When the two channel sections are combined together (Figure E10.4(b)), the radius of gyration about the x-axis is a constant, while the radius of gyration about the Y_c direction depends on the back-to-back distance d. The best design for this combined section is to achieve the same radius of gyration about both the x and Y_c directions, from which distance d can be determined. Between the two connection points, the best design is to make sure that the two channels do not buckle individually and behave as a composite unit before the critical buckling load of the full length column has been reached. This can be used as the requirement to determine h.*

The section properties of C130 × 10 are as follows:

$$A = 12.71\,cm^2, \quad I_x = 312\,cm^4, \quad I_y = 19.9\,cm^4$$

$$r_x = 4.95\,cm, \quad r_y = 1.25\,cm, \quad x_0 = 1.23\,cm$$

(a) For the single section, since $r_y < r_x$ and

$$\frac{\lambda_y L}{r_y} = \frac{1 \times 12 \times 10^2\,cm}{1.25\,cm} = 960$$

$$\geq \pi \sqrt{\frac{E}{\sigma_e}} = \pi \sqrt{\frac{200 \times 10^9\,Pa}{200 \times 10^6\,Pa}} = 99.35$$

the column will fail due to buckling at:

$$P_{cr} = \frac{\pi^2 E I_y}{(\lambda_y L)^2} = \frac{\pi^2 \times 200 \times 10^9\,Pa \times 19.9 \times 10^{-8}\,m^4}{(1 \times 12\,m)^2} = 2.73\,kN$$

(b) For the combined section, the second moment of area about the x-axis is twice that of the single section. The second moment of area about the Y_c-axis varies depending on the back-to-back distance d. For a single section, the second moment of area about the Y_c-axis (Figure E10.4(b)) is calculated by the parallel axis theorem (Equation (5.4)).

$$I_{Y_c} = I_y + A(x_0 + \frac{d}{2})^2 = 19.9\,\text{cm}^4 + 12.71\,\text{cm}^2 \times \left(1.23\,\text{cm} + \frac{d}{2}\right)^2$$

To achieve an equal radius of gyration about both the x- and the Y_c-axis requires

$$2 \times I_{Y_c} = 2 \times I_x \qquad \text{or}$$

$$19.9\,\text{cm}^4 + 12.71\,\text{cm}^2 \times (1.23\,\text{cm} + \frac{d}{2})^2 = 312\,\text{cm}^4$$

Thus:

$$d = 2 \left(\sqrt{\frac{312 - 19.9}{12.71}} - 1.23 \right) = 7.13\,\text{cm}$$

Within the two connection points of distance h, buckling of individual columns occurs when the compression in the columns reaches a certain level. To prevent this local failure, the minimum requirement is that the critical local buckling load (local buckling mode) is greater than or at least equal to half of the critical buckling load of the entire column (overall buckling mode). Thus the slenderness ratio of a single column between the two connection points must be smaller or at least equal to two times the slenderness ratio of the entire column having the combined section:

$$\frac{\lambda_y h}{r_y} \le 2 \times \frac{\lambda_y L}{2r_{Y_c}} = 2 \times \frac{\lambda_x L}{2r_x} = \frac{\lambda_x L}{r_x}$$

Hence:

$$\frac{1 \times h}{1.25 \times 10^{-2}\,\text{m}} \le \frac{1 \times 12\,\text{m}}{4.95 \times 10^{-2}\,\text{m}}$$

From which:

$$h \le 3.03\,\text{m}$$

With the above-calculated h and d, the slenderness ratio of the composite column for the overall buckling satisfies:

$$\frac{\lambda_x L}{2r_x} = \frac{1 \times 12\,\text{m}}{2 \times 4.95 \times 10^{-2}\,\text{m}} = 121.2 \ge \pi \sqrt{\frac{E}{\sigma_p}}$$

The composite column will fail by buckling, and the critical buckling load for the composite column with the above-calculated d and h is:

$$P_{cr} = \frac{\pi^2 E(2 \times I_x)}{(\lambda_x L)^2} = \frac{\pi^2 \times 200 \times 10^9 \text{ Pa} \times 2 \times 312 \times 10^{-8} \text{ m}^4}{(1 \times 12 \text{ m})^2} = 85.54 \text{ kN}$$

By constructing the composite section, the critical buckling load of the column increases from 2.73 kN to 85.54 kN, an increase of 31 times.

EXAMPLE 10.5
Derive the solution for buckling of columns with general end conditions.

[Solution] *Euler formula (Equation (10.3)) is given on the basis of knowing the effective length of a column. The effective length factors for selected cases are given in Table 10.1. This example shows how these factors can be calculated and how the buckling load of columns with other support conditions can be evaluated. The solution starts with solving general bending equation from Example 10.1. In order to include the solution for columns with general end conditions (each end should have two conditions to describe), a fourth-order differential equation is needed.*

From Example 10.1, the deflection and bending moment for a buckled column is related by:

$$\frac{d^2 y}{dx^2} + k^2 y = 0$$

Differentiating the equation twice yields:

$$\frac{d^4 y}{dx^4} + k^2 \frac{d^2 y}{dy^2} = 0$$

where $k^2 = P/EI$. The solution of this fourth-order differential equation is:

$$y = C_1 \sin kx + C_2 \cos kx + C_3 x + C_4$$

The four unknown constants are to be determined through the introduction of support conditions. The derivatives of the above solution are given as:

$$\frac{dy}{dx} = C_1 k \cos kx - C_2 k \sin kx + C_3$$

$$\frac{d^2 y}{dx^2} = -C_1 k^2 \sin kx - C_2 k^2 \cos kx$$

$$\frac{d^3 y}{dx^3} = -C_1 k^3 \cos kx + C_2 k^3 \sin kx$$

The above derivatives are related to the rotation, bending moment and shear force (Equations (6.1) and (6.2)) of the column. Introducing appropriate deflection, rotation, bending moment and shear force conditions at both ends of the column leads to a solution of k, from which the buckling loads can be calculated.

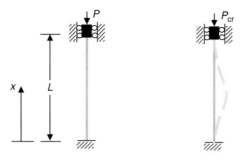

Figure E10.5

For a column with fixed ends (Figure E10.5), e.g., the end conditions are:

$$\text{at } x = 0, \quad y(0) = \frac{dy}{dx}(0) = 0$$

$$\text{at } x = L, \quad y(L) = \frac{dy}{dx}(L) = 0$$

Introducing the above conditions into the solutions yields:

$$y(0) = 0: \quad C_2 + C_4 = 0$$

$$\frac{dy}{dx}(0) = 0: \quad C_1 k + C_3 = 0$$

$$y(L) = 0: \quad C_1 \sin kL + C_2 \cos kL + C_3 L + C_4 = 0$$

$$\frac{dy}{dx}(L) = 0 \quad C_1 k \cos kL - C_2 k \sin kL + C_3 = 0$$

To satisfy this set of equations, C_1, C_2, C_3 and C_4 could take zero, which are trivial solutions that show no deflection of the column. However, when the column buckles, at least one of the four constants must not be zero. This requires:

$$\det \begin{vmatrix} 0 & 1 & 0 & 1 \\ k & 0 & 1 & 0 \\ \sin kL & \cos kL & L & 1 \\ k \cos kL & -k \sin kL & 1 & 0 \end{vmatrix} = 0$$

The evaluation of this determinant yields:

$$2 \cos kL + kL \sin kL - 2 = 0$$

The minimum value of kL satisfying the about equation is $kL = 2\pi$. Thus:

$$\sqrt{\frac{P}{EI}} L = 2\pi$$

$$\text{or } P = \frac{4\pi^2 EI}{L^2} = \frac{\pi^2 EI}{(0.5L)^2}$$

The effective length factor is 0.5, which is exactly the same as the one shown in Table 10.1.

10.5   Conceptual questions

1. Define the term 'column' and 'slenderness' and explain the term 'slenderness ratio' as applied to columns.
2. Explain how the Euler formula for pin-ended columns can be modified for columns having one or both ends fixed.
3. Which of the following statement defines the term 'critical buckling load'?

    (a) The stress on the cross-section due to the critical load equals the proportional limit of the material.
    (b) The stress on the cross-section due to the critical load equals the strength of the material.
    (c) The maximum compressive axial force that can be applied to a column before any lateral deflection occurs.
    (d) The minimum compressive axial force that can be applied to a column before any lateral deflection occurs.
    (e) The axial compressive axial force that causes material failure of a column.

4. Why does the Euler formula become unsuitable at certain values of slenderness ratio?
5. Defined the term 'local buckling'.
6. A column with the following cross-sections (Figure Q10.6) is under axial compression; in which direction the column may buckle?

Figure Q10.6

7. If two columns made of the same material have the same length, cross-sectional area and end supports, are the critical buckling loads of the two columns the same?
8. The long bar shown in Figure Q10.8 is fixed at one end and elastically restrained at the other end. Which of the following evaluations of its effective length factor is correct?

Figure Q10.8

    (a) $\lambda < 0.5$;   (b) $0.5 < \lambda < 0.7$;   (c) $0.7 < \lambda < 2$;   (d) $\lambda > 2$
9. Two compression members are made of the same material and have equal cross-sectional area. Both members have also the same end supports. If the members have square and circular sections, respectively, which of the member has greater critical buckling load?
10. Consider the columns shown in Figure Q10.10. All columns are made of the same material and have the same slenderness ratio. The columns are all pinned at both ends, two of which have also intermediate supports (bracing). If the critical buckling load of column (a) is P_{cr}, calculate the critical buckling loads of the remaining columns.

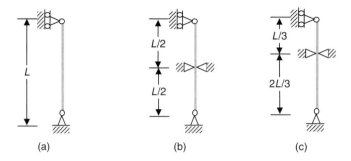

(a)   (b)   (c)

Figure Q10.10

10.6 Mini test

Problem 10.1:   Describe the factors that affect critical buckling load of a column and how stability capacity of the column can be enhanced.

Problem 10.2:   A compressive axial force is applied to two steel rods that have the same material, support conditions and cross-sectional area. If the cross-sections of the two rods are, respectively, solid circular and hollow circular, which rod will buckle at a lower force and why?

Problem 10.3:   A bar of solid circular cross-section has built-in end conditions at both ends and a cross-sectional area of $2\,cm^2$. The bar is just sufficient to support an axial load of 20 kN before buckling. If one end of the bar is now set free from any constraint, it is obvious that the bar has a lower critical buckling load (Figure P10.3). The load-carrying capacity, however, can be maintained without consuming extra material but by making the bar hollow to increase the second moment of cross-sectional area. Determine the external diameter of such a hollow bar.

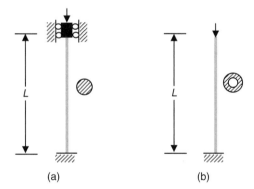

(a)   (b)

Figure P10.3

Problem 10.4:   The pin-joined frame shown in Figure P10.4 carries a downward load P at C. Assuming that buckling can only occur in the plane of the frame, determine the value of P that will cause instability. Both members have a square section of 50 mm × 50 mm. Take $E = 70\,GPa$ for the material.

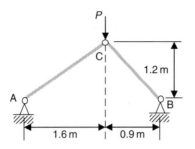

Figure P10.4

Problem 10.5:   A column is built up of two channels with two thin plates bolted to the flanges as shown in Figure P10.5. Calculate the required distance between the backs of the two channels in order to achieve an equal buckling resistance about both the X- and the Y-axis. If the column is 1 m long and fixed at both ends, determine its critical buckling load. Take $E = 80$ GPa.

Figure P10.5

11 Energy method

In the previous chapters, the structural and stress problems were solved on the basis of static equilibrium by considering the relationship between the internal forces (stresses) and the externally applied loads. The same problems can also be solved on the basis of the principle of conservation of energy by considering the energy built up within a body. The energy is stored due to the deformation in relation to the work done by the externally applied loads. In general, the principle of conservation of energy in structural and stress analysis establishes the relationships between stresses, strains or deformations, material properties and external loadings in the form of energy or work done by internal and external forces. The basic concepts of work and energy are as follows:

- Work is defined as the product of a force and the distance in the direction of the force.
- Energy is defined as the capacity to do work.

Unlike stresses, strains or displacement, energy or work is a scalar quantity. Simple application of an energy method is to equalize the work done by external loads and the energy stored in a deformed body, while the most powerful method that can effectively solve a complex structural and stress problem is based on the principle of virtual work.

11.1 Work and strain energy

11.1.1 Work done by a force

In order to accomplish work on an object there must be a force exerted on the object and it must move in the direction of the force. The unit of work is, for example, N m, and is also called Joules.

In Figure 11.1, the work done by the force, F, to move the mass by a distance, d, is

$$\text{Work} = F \times \cos\theta \times d \qquad (11.1)$$

Figure 11.1

Table 11.1 Strain energy

Axial deformation	Bending	Torsion
Internal axial force: $N(x)$ Young's modulus: E Cross-sectional area: A Member length: L	Bending moment: $M(x)$ Young's modulus: E Second moment of area: $I(x)$ Member length: L	Twist moment: $T(x)$ Shear modulus: G Second polar moment of area: $J(x)$ Member length: L
$U = \dfrac{1}{2}\int_0^L \dfrac{N^2(x)}{EA}dx$	$U = \dfrac{1}{2}\int_0^L \dfrac{M^2(x)}{EI}dx$	$U = \dfrac{1}{2}\int_0^L \dfrac{T^2(x)}{GJ}dx$

11.1.2 Strain energy

In a solid deformable body, the work done by stresses on their associated deformation (strains) is defined as strain energy. In general, strain energy is computed as:

$$u = \int_0^{\varepsilon} \sigma d\varepsilon \tag{11.2a}$$

and

$$U = \int_{Vol} u dV \tag{11.2b}$$

where u denotes strain energy per unit volume and U is the total strain energy stored in a body. The strain energies due to different types of deformation are listed in Table 11.1.

If a member is subjected to a combined action of axial force, bending moment and torque, the strain energy is:

$$U = \frac{1}{2}\int_0^L \frac{N^2(x)}{EA}dx + \frac{1}{2}\int_0^L \frac{M^2(x)}{EI}dx + \frac{1}{2}\int_0^L \frac{T^2(x)}{GJ}dx \tag{11.3}$$

11.2 Solutions based on energy method

The linearly elastic system shown in Figure 11.2 is subjected to a set of point loads. The following theorems establish the relationship between the applied forces and the displacements in the directions of the forces.

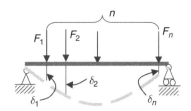

Figure 11.2

11.2.1 Castigliano's first theorem

> If the strain energy of an elastic structural system is expressed in terms of n independent displacements, $\delta_1, \delta_2, \ldots, \delta_n$, associated with a system of prescribed forces $F_1, F_2, \ldots, F_n$, the first partial derivative of the energy with respect to any of these displacements δ_i, is equal to force, F_i, at point i in the direction of δ_i.

The mathematical expression of the theorem is:

$$\frac{\partial U}{\partial \delta_i} = F_i \qquad (11.4)$$

In order to use Castigliano's first theorem, the strain energy must be expressed in terms of displacements δ_i $(i = 1, 2, \ldots, n)$.

11.2.2 Castigliano's second theorem

> If the strain energy of a linear elastic structural system is expressed in terms of n independent forces , $F_1, F_2, \ldots, F_n$, associated with a system of displacements, $\delta_1, \delta_2, \ldots, \delta_n$, the first partial derivative of the energy with respect to any of these forces, F_i, is equal to displacement, δ_i, at point i in the direction of F_i.

The mathematical expression of the theorem is:

$$\frac{\partial U}{\partial F_i} = \delta_i \qquad (11.5)$$

In order to use Castigliano's second theorem, the strain energy must be expressed in terms of forces F_i $(i = 1, 2, \ldots, n)$.

For the energy expressed in the form of Equation (11.3):

$$\delta_i = \frac{\partial U}{\partial F_i} = \int_0^L \frac{N(x)}{EA} \frac{\partial N(x)}{\partial F_i} dx + \int_0^L \frac{M(x)}{EI} \frac{\partial M(x)}{\partial F_i} dx + \int_0^L \frac{T(x)}{GJ} \frac{\partial T(x)}{\partial F_i} dx \qquad (11.6)$$

In the two theorems (Equations (11.4) and (11.5)), the general force, F_i, can be a moment for which the associated displacement is the rotation at the same point.

11.3 Virtual work and the principle of virtual work

11.3.1 Virtual work

A force, F, which may be real or imaginary and acts on an object that is in equilibrium under a given system of loads, is said to do virtual work when the object is imagined to undergo a real or imaginary displacement in the direction of the force. Since the force and/or the displacements are not necessarily real, the work done is called *virtual work*.

Figure 11.3

Virtual work is classified as follows:

- External virtual work if the work is done by real or imaginary externally applied forces on a unrelated real or imaginary displacements of a system.
- Internal virtual work if the work is done by real or imaginary stresses on unrelated real or imaginary strains of a system.

To accomplish virtual work, the virtual displacement or deformation can be any unrelated deformation, but must satisfy the support conditions or boundary constraints of the system. For example, Figure 11.3 shows a beam under two separate load-displacement systems. Case (a) shows the beam's real deformation under the action of the loads shown. Case (b) shows the same beam that undergoes a real deformation under the action of a force, F_b.

If the deformation of Case (a) is taken as the virtual deformation of Case (b), additional to the real deformation Case (a) has already had, the external virtual work done by F_b is:

$$\delta W_e = F_b \times d_a$$

The internal virtual work done by the stresses caused by F_b in Case (b) on the virtual strains (real strain of Case (a)) is:

$$\delta W_i = \int_L \frac{M_b(x)M_a(x)}{EI}dx$$

where $M_a(x)$ and $M_b(x)$ are, respectively, the bending moments in Case (a) and Case (b).

11.3.2 The principle of virtual work

The principle of virtual work is one of the most effective methods for calculating deflections (deformation). The principle of virtual work states as follows: if an elastic body under a system of external forces is given a small virtual displacement, then the increase in work done by the external forces is equal to the increase in strain energy stored.

In Figure 11.3, by taking the deformation of Case (a) as the virtual deformation of Case (b), the principle of virtual work yields:

$$\delta W_e = \delta W_i$$

or

$$F_b \times d_a = \int_L \frac{M_a(x)M_b(x)}{EI}dx \tag{11.7}$$

In particular, if F_b is chosen as a unit force, the above equation yields the mid-span deflection of the beam under the loads shown in Case (a), that is:

$$d_a = \int_L \frac{M_a(x)M_b(x)}{EI}dx \tag{11.8}$$

Equation (11.8) is called solution of the *unit load method*. For a general case involving axial deformation, bending and torsion, the solution is:

$$\Delta_P = \int_0^L \frac{N_a(x)N_b(x)}{EA}dx + \int_0^L \frac{M_a(x)M_b(x)}{EI}dx + \int_0^L \frac{T_a(x)T_b(x)}{GJ}dx \tag{11.9}$$

where

$N_a(x)$, $M_a(x)$ and $T_a(x)$ are the real internal forces of a system subject to externally applied loads.

$N_b(x)$, $M_b(x)$ and $T_b(x)$ are the internal forces if the same system subject to a unit force (moment) at a particular point and in a particular direction.

Δ_P is the displacement/rotation of the system at the location, subject to the real external loads, where the unit force (moment) is applied.

Note: The displacement/rotation is in the same direction of the applied unit force/moment if the computed displacement/rotation is positive. Otherwise, it is in the opposite direction of the applied unit load.

11.3.3   Deflection of a truss system

Within the members of a truss system (pin-joined frame), there is no bending and twist moments and the axial forces are constant along the length of each member. Thus, for a truss comprising n members, Equation (11.9) is reduced to:

$$\Delta_P = \int_0^L \frac{N_a(x)N_b(x)}{EA}dx = \sum_{i=1}^n \frac{N_a^{(i)}N_b^{(i)}}{E_iA_i}L_i \tag{11.10}$$

where $N_a^{(i)}$ and $N_b^{(i)}$, E_i, A_i and L_i are the respective forces, Young's modulus, cross-sectional area and length of the *i*th member. They are usually all constant within a member.

Application of the unit load method Equation (11.10) to find deflection of a truss system, for example the vertical deflection at joint A of the truss shown in Figure 11.4, follows the procedure below:

A

Figure 11.4

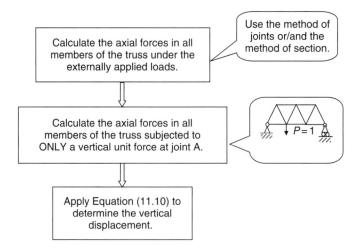

11.4   Key points review

- Any force will do work if it is associated with a deformation (displacement).
- Work is a scalar quality.
- Energy can be stored due to tension, compression, shearing, bending and twisting.
- Strain energy can be related to the change in dimension of body.
- For a material subject to externally applied loads, the work done by the applied loads must equal the strain energy stored in the material.
- Virtual work can be done by real force on imaginary (virtual) displacement or imaginary (virtual) force on real displacement.
- A imaginary (virtual) deformation of a system can be any possible deformation of the system that satisfies the support conditions, including the real deformation of the system.
- Castiglianos's theorems provide relationships between a particular deformation and a particular force at a point.
- The unit load method provides a convenient tool for computing displacements in structural and stress analysis. It is applicable for both linear and nonlinear materials.
- A displacement obtained from applying the unit load method can be negative. In such a case the displacement is in the opposite direction of the applied unit force.

11.5   Examples

EXAMPLE 11.1
The indeterminate truss shown in Figure E11.1 consists of three members having the same value of EA. The truss is subjected to a downwards force F as shown. Determine the axial forces in the three members by using Castigliano's first theorem.

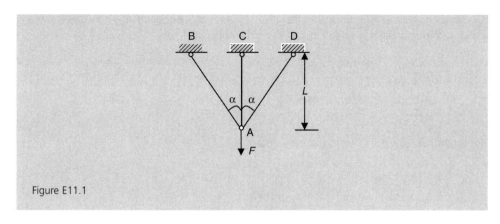

Figure E11.1

[Solution] *In order to use Castigliano's first theorem, the strain energy must be expressed in terms of displacement. For such a symmetric system, we assume that joint A has a small vertical displacement Δ, which is the elongation of member AC. The elongation of members AB and AD can be easily determined from geometrical relations. The elongations are used to calculate the strain energy that can be subsequently used, along with the equilibrium condition, to compute the axial forces within each of the members.*

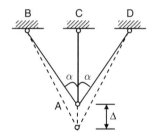

Assume that the elongation of AC, AB and AD are, respectively, Δ, Δ_{AB} and Δ_{AD}. For element AC under axial tension, applying Hooke's law yields:

$$\sigma = E\varepsilon$$

or

$$\frac{N_{AC}}{A} = E\frac{\Delta}{L}$$

then:

$$N_{AC} = EA\frac{\Delta}{L}$$

The strain energy of this member is (Table 11.1):

$$U_{AC} = \frac{1}{2}\int_0^L \frac{N^2(x)}{EA}dx = \frac{1}{2}EA\left(\frac{\Delta}{L}\right)^2 \times L = \frac{EA\Delta^2}{2L}$$

For members AB and AD, since their length is $L/\cos\alpha$ and elongations are, respectively, Δ_{AB} and Δ_{AD}, the strain energy stored in these two members are:

$$U_{AB} = \frac{EA(\Delta_{AB})^2}{2(L/\cos\alpha)} = \frac{EA(\Delta_{AB})^2 \cos\alpha}{2L}$$

$$U_{AD} = \frac{EA(\Delta_{AD})^2}{2(L/\cos\alpha)} = \frac{EA(\Delta_{AD})^2 \cos\alpha}{2L}$$

The total strain energy stored in the truss system is:

$$U = U_{AB} + U_{AC} + U_{AD}$$

$$= \frac{EA(\Delta_{AB})^2 \cos\alpha}{2L} + \frac{EA\Delta^2}{2L} + \frac{EA(\Delta_{AD})^2 \cos\alpha}{2L}$$

By Castigliano's first theorem, the axial forces in these members are, respectively:

$$N_{AC} = \frac{\partial U}{\partial\Delta} = \frac{EA\Delta}{L}$$

$$N_{AB} = \frac{\partial U}{\partial\Delta_{AB}} = \frac{EA\Delta_{AB}}{L}\cos\alpha$$

$$N_{AD} = \frac{\partial U}{\partial\Delta_{AD}} = \frac{EA\Delta_{AD}}{L}\cos\alpha$$

From the geometry, the elongation of AC, AB and AD are, respectively, Δ, $\Delta_{AB} = \Delta\cos\alpha$ and $\Delta_{AD} = \Delta\cos\alpha$. Thus:

$$N_{AB} = \frac{EA\Delta_{AB}}{L}\cos\alpha = \frac{EA\Delta}{L}\cos^2\alpha$$

$$N_{AD} = \frac{EA\Delta_{AD}}{L}\cos\alpha = \frac{EA\Delta}{L}\cos^2\alpha$$

Applying the method of joint to A yields:

$$N_{AC} + N_{AB}\cos\alpha + N_{AD}\cos\alpha = F$$

$$\frac{EA\Delta}{L} + \frac{EA\Delta}{L}\cos^3\alpha + \frac{EA\Delta}{L}\cos^3\alpha = F$$

$$\Delta = \frac{F}{\dfrac{EA}{L} + \dfrac{EA}{L}\cos^3\alpha + \dfrac{EA}{L}\cos^3\alpha}$$

$$= \frac{F}{\dfrac{EA}{L} + 2\dfrac{EA}{L}\cos^3\alpha}$$

Substituting the above Δ into the expressions of the three member forces yields:

$$N_{AC} = \frac{EA\Delta}{L} = \frac{F}{1 + 2\cos^3\alpha}$$

$$N_{AB} = N_{AD} = \frac{EA\Delta}{L}\cos^2\alpha = \frac{F\cos^2\alpha}{1 + 2\cos^3\alpha}$$

EXAMPLE 11.2
A cantilever beam supports a uniformly distributed load, q. Use Castigliano's second theorem to determine the deflections at points A and B (E and I are constant).

Figure E11.2

[Solution] *To use Castigliano's second theorem, imaginary forces F_A and F_B are applied at points A and B, respectively, in the calculation of strain energy. To remove them, these imaginary forces will be replaced by zeros after the derivatives have been taken.*

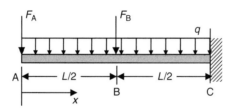

The strain energy for a beam is (Table 11.1):

$$U = \frac{1}{2}\int_0^L \frac{M^2(x)}{EI}dx$$

The bending moment due to the applied distributed load and the two imaginary forces is (see Section 6.2.3):

$$M(x) = -F_A\langle x - 0\rangle - F_B\left\langle x - \frac{L}{2}\right\rangle - \frac{1}{2}q\langle x - 0\rangle^2$$

$$= -F_A x - F_B\left\langle x - \frac{L}{2}\right\rangle - \frac{1}{2}qx^2$$

By Castigliano's second theorem (Equation (11.6)), the vertical deflections at A and B are, respectively:

$$\delta_A = \frac{\partial U}{\partial F_A} = \int_0^L \frac{M(x)}{EI} \frac{\partial M(x)}{\partial F_A} dx$$

> Introduce $F_A = F_B = 0$ after the derivative.

$$= \int_0^L \frac{-P_A x - P_B \langle x - L/2 \rangle - qx^2/2}{EI}(-x)dx$$

$$= \int_0^L \frac{qx^2/2}{EI} x\, dx = \frac{qL^4}{8EI}$$

$$\delta_B = \frac{\partial U}{\partial F_B} = \int_0^L \frac{M(x)}{EI} \frac{\partial M(x)}{\partial F_B} dx$$

> Introduce $F_A = F_B = 0$ after the derivative.

$$= \int_0^L \frac{-P_A x - P_B \langle x - L/2 \rangle - qx^2/2}{EI} \times (-1)\left\langle x - \frac{L}{2} \right\rangle dx$$

$$= \int_0^L \frac{qx^2/2}{EI} \langle x - L/2 \rangle\, dx$$

$$= \int_{L/2}^L \frac{qx^2/2}{EI}(x - L/2)dx = \frac{17qL^4}{384}$$

The above deflections can be verified by the formulas in Table 6.2.

EXAMPLE 11.3

Use the unit load method to find the deflections at A and B of the beam shown in Figure E11.3.

Figure E11.3

[Solution] *To apply Equation (11.9) to solve this problem, Case (a) is set as the beam subject to the uniformly distributed load and Case (b) is set as the beam subject to a unit downward force applied at points A and B, respectively.*

The bending moments of Cases (a) and (b) can be calculated and introduced into Equation (11.9) to compute the deflections.

The bending moment of Case (a) is:

$$M_a(x) = -\frac{qx^2}{2}$$

The bending moments of Case (b) are, respectively:

$$M_b(x) = -F_A \times x = -x$$

when only F_A is applied, and

$$M_b(x) = 0, \qquad\qquad\qquad\qquad 0 \le x \le L/2$$
$$M_b(x) = -F_B \times (x - L/2) = -(x - L/2), \qquad L/2 \le x \le L$$

when only F_B is applied. Thus:

$$\delta_A = \int_L \frac{M_a(x)M_b(x)}{EI}\,dx = \int_L \frac{(-qx^2/2)(-x)}{EI}\,dx = \frac{qL^4}{8EI}$$

$$\delta_B = \int_L \frac{M_a(x)M_b(x)}{EI}\,dx = \int_0^{L/2} \frac{(-qx^2/2)(0)}{EI}\,dx + \int_{L/2}^L \frac{(-qx^2/2)(-x+L/2)}{EI}\,dx$$

$$= \frac{17qL^4}{384EI}$$

The results are identical to the solutions from Example 11.2.

EXAMPLE 11.4
The plane frame structure is loaded as shown in Figure E11.4. Determine the horizontal displacement, the vertical deflection and the angle of rotation of the section at C. The stiffness of the two members, EI, is constant. Ignore axial and shear deformation of the members.

Figure E11.4

[Solution] *The horizontal displacement, the vertical deflection and angle of rotation at C can be determined by applying a unit horizontal force, a vertical force and a unit bending moment, respectively, at C. Since both the axial and the shear deformation are ignored, only the strain energy due to bending is required when applying the unit load method. In Equation (11.8), the system shown in Figure E11.4 is taken as Case (a), and Case (b) is taken as follows:*

Case (b) for
horizontal
displacement

Case (b) for
vertical
deflection

Case (b) for
rotation

Since the frame consists of members of different orientations, local coordinates, x_1 for the column and x_2 for the beam, are set to simplify calculation of the bending moment.

	Bending moment in the column from A to B $0 \le x_1 \le a$	Bending moment in the beam from C to B $0 \le x_2 \le a$
Case (a) Figure E11.4	$M_a(x) = -\dfrac{qa^2}{2}$	$M_a(x) = -\dfrac{qx_2^2}{2}$
Case (b) Horizontal displacement at C	$M_b(x) = -P_b^H(a-x_1) = -(a-x_1)$	0
Case (b) deflection at C	$M_b(x) = -P_b^V a = -a$	$M_b(x) = -P_b^V x_2 = -x_2$
Case (b) rotation at C	$M_b(x) = -M_b = -1$	$M_b(x) = -M_b = -1$

Introducing the bending moments from the above table into Equation (11.8) yields the following. The horizontal displacement at C is:

$$d_C^V = \int_L \frac{M_a(x)M_b(x)}{EI}\,dx = \int_{CB} \frac{(-qx_2^2/2)\times(0)}{EI}\,dx_2 + \int_{AB} \frac{(-qa^2/2)\times(x_1-a)}{EI}\,dx_1$$

$$= \frac{qa^4}{4EI}$$

The vertical deflection at C is:

$$d_C^V = \int_L \frac{M_a(x)M_b(x)}{EI}\,dx = \int_{CB} \frac{(-qx_2^2/2)\times(-x_2)}{EI}\,dx_2 + \int_{AB} \frac{(-qa^2/2)\times(-a)}{EI}\,dx_1$$

$$= \frac{5qa^4}{8EI}$$

The angle of rotation at C is:

$$\theta_C = \int_L \frac{M_a(x)M_b(x)}{EI}\,dx = \int_{CB} \frac{(-qx_2^2/2)\times(-1)}{EI}\,dx_2 + \int_{AB} \frac{(-qa^2/2)\times(-1)}{EI}\,dx_1$$

$$= \frac{2qa^3}{3EI}$$

EXAMPLE 11.5

An aluminium wire 7.5 m in length with a cross-sectional area of 100 mm² is stretched between a fixed pin and the free end of the cantilever as shown in Figure E11.5. The beam is subjected to a uniformly distributed load of 12 kN/m. The Young's modulus and moment of initial of the beam are, respectively, 200 GPa and 20×10^6 mm⁴. Determine the force in the wire.

Figure E11.5

[Solution] *This is a statically indeterminate structure of first order. The joint at A can be released and replaced by an unknown axial force F_A acting at A. The strain energy of the system can then be calculated in terms of the applied distributed load and the unknown axial force. From the Castigliano's second theorem (Equation (11.6)), the derivative of the strain energy with respect to F_A yields the displacement of A in the vertical direction. This vertical displacement must be zero since the point is pinned to the ceiling, which provides the following equation for the solution of the unknown axial force, F_A.*

$$\frac{\partial U}{\partial F_A} = 0$$

The axial force in the wire is F_A and the strain energy in AB is:

$$U_{AB} = \frac{1}{2}\int_A^B \frac{F_A^2}{EA}dx = \frac{1}{2}\frac{F_A^2 L_{AB}}{EA}$$

The bending moment in the beam is $F_A x - \frac{1}{2}qx^2$ and the strain energy in BC is:

$$U_{BC} = \frac{1}{2} \int_B^C \frac{M(x)^2}{EI} dx = \frac{1}{2} \int_B^C \frac{(F_A x - qx^2/2)^2}{EI} dx$$

Then $U = U_{AB} + U_{BC}$ and

$$\frac{\partial U}{\partial F_A} = \frac{\partial U_{AB}}{\partial F_A} + \frac{\partial U_{BC}}{\partial F_A} = \frac{F_A L_{AB}}{EA} + \int_B^C \frac{(F_A - qx^2/2)}{EI} x dx$$

$$= \frac{F_A L_{AB}}{EA} + \frac{1}{EI}\left[\frac{F_A L_{BC}^2}{2} - \frac{qL_{BC}^4}{8}\right] = 0$$

So:

$$\frac{F_A \times 7.5\,m}{70 \times 10^9\,N/m^2 \times 100 \times 10^{-6}\,m^2} + \frac{F_A \times (3\,m)^2/2}{200 \times 10^9\,N/m^2 \times 20 \times 10^{-6}\,m^4}$$

$$= \frac{12 \times 10^3\,N/m \times (3\,m)^4/8}{200 \times 10^9\,N/m^2 \times 20 \times 10^{-6}\,m^4}$$

which yields:

$$F_A = 9.145\,kN$$

The axial force in the wire is 9.145 kN.

EXAMPLE 11.6
To determine the deflection at A of the beam (Figure E11.6) loaded with a point force P and a bending moment PL, the following solution is obtained by using Castigliano's second theorem. Is the solution correct and why?

Figure E11.6

The bending moment of the beam is:

$$M(x) = PL - Px = P(L - x)$$

From Equation (10.6)

$$\delta_B = \frac{\partial U}{\partial P} = \int_0^L \frac{M(x)}{EI} \frac{\partial M(x)}{\partial P} dx = \int_0^L \frac{P(L-x)}{EI}(L-x)dx$$

$$= \frac{PL^3}{3EI}$$

[Solution] *The question tests your understanding of Castigliano's second theorem. The strain energy in Equation (11.6) must be expressed in terms of the applied loads that must be considered as independent forces. In this question the applied bending moment is related to the applied point force by P. Thus the derivative with respective to P is taken in relation to not just the force, but also the applied moment. The solution shown above is, therefore, not correct. The correct solution can be obtained by expressing the strain energy in terms of the force P and a bending moment, M_P, that is considered completely independent of P. After the derivatives with respect to P is taken, the bending moment is replaced by PL to obtain the deflection.*

The bending moment of the beam is:

$$M(x) = M_P - Px$$

From Equation (11.6):

$$\delta_B = \frac{\partial U}{\partial P} = \int_0^L \frac{M(x)}{EI} \frac{\partial M(x)}{\partial P} dx = \int_0^L \frac{M_P - Px}{EI}(-x)dx$$

$$= \int_0^L \frac{PL - Px}{EI}(-x)dx = -\frac{PL^3}{6EI}$$

> M_P is replaced by PL here.

Because the final solution is negative, the deflection at A is in the opposite direction of the applied force *P*.

EXAMPLE 11.7

Find the vertical deflection of point E in the pin-jointed steel truss shown in Figure E11.7 due to the applied loads at B and F. EA is constant for all members.

Case (a)

Figure E11.7

[Solution] *This is a typical example showing how the unit load method can be applied to find deflection of a truss system. Since this is a pin-jointed structure subject to loads applied through joints only, within each member, only a constant axial force exists and Equation (11.10) applies. Thus, the system shown in Figure E11.7 is taken as Case (a) and the following system is taken as Case (b), where an imaginary unit force is applied vertically at E.*

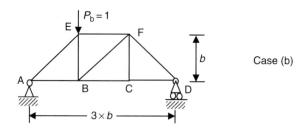

Case (b)

To use Equation (11.10), the axial forces of all the members for Cases (a) and (b) must be calculated first. This can be easily done by the method of joint or/and the method of section.

The calculation of the axial forces and the deflection by Equation (11.10) can be presented in the following tabular form.

Member	Length	N_a Case (a)	N_b Case (b)	$N_a \times N_b \times L$
AB	b	$\dfrac{4P}{3}$	$\dfrac{2}{3}$	$\dfrac{8Pb}{9}$
AE	$\sqrt{2}b$	$-\dfrac{4\sqrt{2}P}{3}$	$-\dfrac{2\sqrt{2}}{3}$	$\dfrac{16\sqrt{2}Pb}{9}$
BC	b	$\dfrac{5P}{3}$	$\dfrac{1}{3}$	$\dfrac{5Pb}{9}$
BF	$\sqrt{2}b$	$-\dfrac{\sqrt{2}P}{3}$	$\dfrac{\sqrt{2}}{3}$	$-\dfrac{2\sqrt{2}Pb}{9}$
BE	b	$\dfrac{4P}{3}$	$-\dfrac{1}{3}$	$-\dfrac{4Pb}{9}$
CD	b	$\dfrac{5P}{3}$	$\dfrac{1}{3}$	$\dfrac{5Pb}{9}$
CF	b	0	0	0
DF	$\sqrt{2}b$	$-\dfrac{5\sqrt{2}P}{3}$	$-\dfrac{\sqrt{2}}{3}$	$\dfrac{10\sqrt{2}Pb}{9}$
EF	b	$-\dfrac{4P}{3}$	$-\dfrac{2}{3}$	$\dfrac{8Pb}{9}$
$\displaystyle\sum_{\text{all members}} \dfrac{N_a N_b L}{EA}$				$6.22\,\dfrac{Pb}{EA}$

The vertical deflection at point E is 6.22 *Pb/EA* downwards.

EXAMPLE 11.8
Consider the same truss system shown in Figure E11.7. Calculate the increase of the distance between points C and E.

Figure E11.8

[Solution] *This question asks for relative displacement of points C and E. Instead of applying unit load at C and E and carrying out respective calculations for Case (b), a pair of unit loads is applied simultaneously as shown in Figure E11.8. After calculating the axial forces of Figure E11.8 and following the same procedure of Example 11.7, the relative displacement of C and E, that is, the increase of distance between the two points, is obtained.*

Replacing the axial forces in the column of N_b (in Example 11.7) by the respective axial forces calculated from Figure E11.8 yields the following:

Member	Length	N_a	N_b	$N_a \times N_b \times L$
AB	b	$\dfrac{4P}{3}$	0	0
AE	$\sqrt{2}b$	$-\dfrac{4\sqrt{2}P}{3}$	0	0
BC	b	$\dfrac{5P}{3}$	$\dfrac{\sqrt{2}}{2}$	$\dfrac{5\sqrt{2}Pb}{6}$
BF	$\sqrt{2}b$	$-\dfrac{\sqrt{2}P}{3}$	-1	$\dfrac{2Pb}{3}$
BE	b	$\dfrac{4P}{3}$	$\dfrac{\sqrt{2}}{2}$	$\dfrac{2\sqrt{2}Pb}{3}$
CD	b	$\dfrac{5P}{3}$	0	0
CF	b	0	$\dfrac{\sqrt{2}}{2}$	0
DF	$\sqrt{2}b$	$-\dfrac{5\sqrt{2}P}{3}$	0	0
EF	b	$-\dfrac{4P}{3}$	$\dfrac{\sqrt{2}}{2}$	$-\dfrac{2\sqrt{2}Pb}{3}$
$\displaystyle\sum_{\text{all members}} \dfrac{N_a N_b L}{EA}$				$1.85\dfrac{Pb}{EA}$

The distance between C and E is increased by 1.85 Pb/EA.

11.6 Conceptual questions

1. Define 'strain energy' and derive a formula for it in the case of a uniform bar in tension.
2. Can strain energy be negative?
3. Can virtual strain energy be negative?
4. When a linearly elastic structure is subjected to more than one load, can the strain energy stored in the structure due to the applied loads be calculated by superposition of the strain energy of the structure under the action of the loads applied individually? Discuss the following two cases.

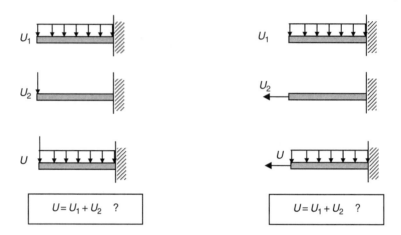

Figure Q3.1

5. Explain how the deflection of beam under a single point load can be found by a strain energy method.
6. State and explain Castigliano's first theorem.
7. State and explain Castigliano's second theorem. How can it be used to determine support reactions of a structure?
8. What are meant by the terms 'virtual force', 'virtual displacement' and 'virtual work'?
9. What is the difference between the work done by a real force and that by a virtual force?
10. Can virtual work be negative?
11. When the unit load method is used to determine the deflection at a point and the calculated deflection is negative, explain why this happens and what the direction of the deflection is.
12. Explain the principle of virtual work and how it can be used in structural and stress analysis.

11.7 Mini test

Problem 11.1:   The beam shown in Figure P11.1 is subjected to a combined action of a force and a moment. Can the strain energy under the combined action be calculated by superposition of the strain energy stored in the beam due to the action of the force and the moment applied separately?

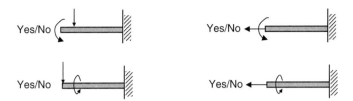

Figure P11.1

Problem 11.2:   The beam shown in Figure P11.2 is subjected to two identical point forces applied at A and B. The strain energy of the beam, U, is expressed in terms of P, and the derivative of U with respect to P, $\partial U/\partial P$, is:

Figure P11.2

A.   the deflection at A
B.   the deflection at B
C.   the average deflection at A and B
D.   the total deflection at A and B.

Problem 11.3:   Determine the deflection at A and rotation at B of the beam shown in Figure P11.3. EI is a constant.

Figure P11.3

Problem 11.4:   Determine the deflection at G of the truss shown in Figure P11.4 using the unit load method. The top and bottom members are made of timber with $E_{tb} = 10\,\text{GPa}$ and $A_{tb} = 200\,\text{cm}^2$. The diagonal members are also made of timber with $E_d = 10\,\text{GPa}$ and $A_d = 80\,\text{cm}^2$. The vertical members are made of steel with $E_{steel} = 20\,\text{GPa}$. The cross-sectional areas of the vertical members are $A_v = 1.13\,\text{cm}^2$ except the central one, whose cross-sectional area is doubled.

Figure P11.4

Problem 11.5:   Find the vertical deflection at point B in the pin-jointed steel truss shown in Figure P11.5 due to the applied load at B. Let $E = 200\,\text{GPa}$. Use Castigliano's second theorem and the unit load method.

Figure P11.5

Index

eBooks – at www.eBookstore.tandf.co.uk

A library at your fingertips!

eBooks are electronic versions of printed books. You can store them on your PC/laptop or browse them online.

They have advantages for anyone needing rapid access to a wide variety of published, copyright information.

eBooks can help your research by enabling you to bookmark chapters, annotate text and use instant searches to find specific words or phrases. Several eBook files would fit on even a small laptop or PDA.

NEW: Save money by eSubscribing: cheap, online access to any eBook for as long as you need it.

Annual subscription packages

We now offer special low-cost bulk subscriptions to packages of eBooks in certain subject areas. These are available to libraries or to individuals.

For more information please contact webmaster.ebooks@tandf.co.uk

We're continually developing the eBook concept, so keep up to date by visiting the website.

www.eBookstore.tandf.co.uk